Felice Fontana, Joseph Skinner

Treatise on the Venom of the Viper

Vol. 1

Felice Fontana, Joseph Skinner

Treatise on the Venom of the Viper
Vol. 1

ISBN/EAN: 9783337375812

Printed in Europe, USA, Canada, Australia, Japan

Cover: Foto ©berggeist007 / pixelio.de

More available books at **www.hansebooks.com**

TREATISE

ON THE

VENOM OF THE VIPER;

ON THE

AMERICAN POISONS;

AND ON THE

CHERRY LAUREL,

AND SOME OTHER

VEGETABLE POISONS.

TO WHICH ARE ANNEXED,

OBSERVATIONS ON THE PRIMITIVE STRUCTURE OF THE ANIMAL BODY; DIFFERENT EXPERIMENTS ON THE REPRODUCTION OF THE NERVES; AND A DE-SCRIPTION OF A NEW CANAL OF THE EYE.

WITH TEN DESCRIPTIVE PLATES.

TRANSLATED FROM THE ORIGINAL FRENCH OF

FELIX FONTANA,

NATURALIST TO HIS ROYAL HIGHNESS THE GRAND DUKE OF TUSCANY, AND DIRECTOR OF HIS CABINET OF NATURAL HISTORY,

BY JOSEPH SKINNER,

NAVY SURGEON, AND MEMBER OF THE CORPORATION OF SURGEONS OF LONDON.

THE SECOND EDITION.

VOL. I.

LONDON:

PRINTED FOR JOHN CUTHELL, NO. 4, MIDDLE ROW, HOLBORN.

M.DCC.XCV.

PREFACE

FRENCH EDITOR.

THE Firſt Part of this Work was publiſhed in Italian in 1765. Monſieur
Darcet, a celebrated Phyſician at Paris,
thought it of ſo much importance, that he
tranſlated it into French a ſhort time after.
Accidental circumſtances delayed the publication of this Tranſlation. The Authour
came to Paris in 1776, and gave M. Darcet
ſeveral ſheets of correċtions and additions,
which were likewiſe tranſlated, and added to
the reſt. On the following year a pamphlet
appeared at Paris on the volatile alkali *(a)*,
in which the nature of the venom of the viper, and the uſe of the volatile alkali againſt
the bite of this animal, were treated of.
Many things were advanced in this pamphlet,
entirely contradiċtory to what our authour had

* Written by M. Sage.

written

written more than ten years before in Italy. He fancied he had been deceived, and fat about making frefh experiments on the fame fubject, determined to correct the tranflation I have fpoken of, before it was given to the publick.

To this new examination, we owe the fecond third and fourth parts of this work, throughout the whole of which the moft delicate experiments are confpicuous. We may fay in the ftricteft fenfe, that thefe parts are entirely new, as well on account of the fubjects of which they treat, as of the difcoveries they contain.

Before François Redi, no one knew in what the venom of the viper confifted. This celebrated naturalift employed the greateft part of his refearches in confuting the errours that prevailed in his time. His work on the venom of the viper is almoft entirely calculated to demonftrate, that the opinions held of this venom were in general falfe: a humiliating reflection to man, who is incapable of attaining the truth, in any other way than by paffing through errour.

We owe to Redi, what has acquired him the greateft reputation, the firft difcovery of
the

the humour that renders the bite of the viper venomous. The experiments he has employed to demonſtrate this diſcovery, are in general very well made, although they have not appeared deciſive to M. Charas, a French chemiſt.

The latter, after having made a great many experiments on the bite of the viper, thought it conclufive that the venom of this animal confiſts in its rage; or rather, that the ſaliva of the viper, exalted by rage, when the creature is furious and bites, becomes venomous and mortal.

Although this opinion is erroneous, it however carries with it marks of probability; ſince it is certain that a viper is more dangerous and deſtructive in proportion as it is more irritated, as will be ſeen in the courſe of this work. This is what engaged our authour to examine the hypotheſis of ſaliva exalted by rage, by deciſive experiments, and the reader may aſſure himſelf, by peruſing the firſt part of this work, that he has been not a little ſucceſsful in this enquiry.

It is certain that François Redi was miſtaken as to the part where the receptacle of the venom is ſeated, and as to the rout the

venom

venom takes, when the viper in biting communicates it to animals. He was of opinion that this humour is fituated in the membrane that covers the canine teeth, and that it glides exteriourly along the tooth, and at the fame time infinuates itfelf into the animals that receive the bite. We fee this errour repeated half a century afterwards in James's Dictionary, the author of which befides adopts all the errours of Mead on the faline nature of the venom. If it was not therefore abfolutely neceffary, it was certainly ufeful, to examine this matter afrefh, and to make it as clear as poffible.

All the other refearches our authour has made, properly belong to him, and we may fay with reafon, that he has begun where others have ended; or, with greater juftice, that the whole of his work is new, and truly original.

It appears to me, that one of the greateft merits of this work confifts, not fo much in the rare and numberlefs difcoveries it contains, as in the luminous method with which the very important enquiries that are introduced in it are treated. If we are aftonifhed at the immenfe number of errours that are

every

every where exploded, we cannot at the fame time help admiring the tracks, hitherto unknown to obfervers, that our authour has beaten, in examining the fubject of poifons.

But that which deferves the greateft confideration, is the very delicate analyfis he has made of the moft obfcure and complicated enquiries; and the fagacity with which he has conceived the experiments, that could not avoid leading him to truth. It is to be wifhed that he may ferve in future as a model to thofe philofophers who enquire into facts, without prepoffeffion or prejudice. How many difputes and opinions will then be terminated! How many truths will be difcovered! How many errours exploded! And how will the number of books be reduced! *The art of interrogating nature by the means of experiment is very delicate. It is in vain that you combine facts, if thefe facts have no affinity with each other; if they prefent themfelves under an equivocal form; if when they are produced by different caufes, you are incapable of affigning and feparating with a certain precifion, the particular effects that have refulted from each of thefe caufes*.*

---

* New Experiments on the Refiftance of Fluids, in the Preliminary Difcourfes of Meffrs. D'Alembert, Condorcet, and Boffut.

To

To judge of what our authour has done in this work, and of what in juſtice belongs to him, the reader ought in the firſt place to examine all the publications of Redi and Mead on the ſame ſubjeçt. I exhort him to do ſo, and this is the greateſt eulogium I can pay to the preſent work. Compariſon, a touchſtone that can never deceive us, is all that I demand, or rather, what juſtice and impartiality require.

Let this work be compared with thoſe on the ſame ſubjeçt that have immortalized Redi and Mead. There will be no difficulty in judging how much it ſurpaſſes theirs, whether we conſider the number of diſcoveries, or the variety and multiplicity of experiments: indeed it will ſoon be ſeen, that there is no room for a compariſon betwixt them.

I regard the having found the venom of the viper to be a gummy ſubſtance, as a real diſcovery. The exiſtence of an animal gum is both important and new.

All that we meet with on the blood, and on the nerves, as they relate to the venom of the viper, is entirely new and original. 'This may be called a giant's ſtride, that clears a new road to freſh diſcoveries.

But

But this is not all the merit of the prefent work. The treatife on the American poifon called Ticunas, and on fome other vegetable poifons; the tracts on the Cherry-laurel; and the experiments on fome other vegetable fubftances, are a new field of enquiry, in which the difcoveries and induftry of our authour are very confpicuous. However, when in confequence of fo many rich difcoveries we think that nothing is any longer hid from us, and that we have at length penetrated into the moft fecret receffes of nature, we meet with labyrinths, out of which it does not feem permitted us to extricate ourfelves. Such are the new and important confequences of the obfervations on the poifon of the Cherry-laurel; a fubject ftill obfcure, but interefting, which will give occafion to the refearches of future obfervers.

The excellent experiments our author has made, relatively to the action of poifons on the nerves, have occafioned him to enrich .ais work with feveral very important enquiries into the ftructure of the nerves themfelves; an obfcure matter, in which nothing was known, and in which it fcarcely feemed allowed to man to make any advances. In

the

the hand of our authour every thing becomes clear, eafy, and fimple. I cannot conceive how this double order of bands, how this fpiral, exteriour, and apparent ftructure, in the nerves, can have efcaped the fight of all our anatomifts; and I regard the certain knowledge we now have of the firft elements of the nerves, as one of the fineft and moft ftriking difcoveries that has ever been made in animal phyficks; a difcovery that has efcaped the fight of the moft fkilful and moft practifed obfervers.

After the obfervations of Lewenhoeck, phyfiologifts and anatomifts were of opinion that a difcovery of the laft divifions of the nerves could never be attained to; but what did not appear poffible then, is now unqueftionably eftablifhed, as any one may affure himfelf, by following the traces of our authour. We have reafon to flatter ourfelves that he will fhortly favour us with his obfervations on the nature and ufes of *the Primitive Nervous Cylinders.* This is all that remains to be known of thefe wonderful organs. He began fome time ago to bufy himfelf in this enquiry, and what ought we

not

not to expect from so exact, and so penetrating
an observer !

· We are at present not only acquainted with
the true structure of the nerves, but are much
better informed than heretofore of that of the
brain.

Our authour has likewise examined the internal structure of the retina—it therefore appears that we have scarcely any thing more to
desire as to that organ.

But his observations do not terminate here.
He has developed with the same success, the
structure of the muscles, and of the tendons.
He finds certain characteristicks to distinguish
these two kinds of substances, as well from
each other, as from the nerves.

The first organical elements of the nerves,
the brain, the muscles, and tendons, being
thus known, our authour passes on to the
discovery of a new and complete system of
transparent, winding, unramified, cylinders,
of a much smaller size than those of the
blood, but which are more extended, and in
a greater number, than the arterial and venous
vessels. He finds them in the whole cellular
substance, a substance which penetrates, and

composes

compofes, all the organs of the animal machine.

Our authour difcovers what he calls winding threads, in hairs, in the nails, in the epidermis, and in the bones. He afterwards gives us a detail of obfervations he made on vegetables, in which it appears, that he traces a fimilar ftructure. He concludes by a feries of excellent obfervations of foffils, on the fubject of which he interfperfes a few doubts, that the reader may not be led away by fimple appearances; and referves his fentiments on this fubject for another work, which he purpofes to publifh under the title of *Microfcopical Obfervations.——Obfervations Microfcopiques.*

He terminates his refearches on the nerves, by relating feveral experiments on the reproduction of them,—a very interefting matter, hitherto unknown to naturalifts, on which he has thrown all poffible light.——

To complete this work, I have judged it neceffary to add a defcription of a new canal of the eye, difcovered by our authour more than eighteen years ago, and which he has never publifhed.—I have taken this defcription from a letter dated from London, which

he

he wrote at the end of the year 1779, to Mr. Adolphus Murray, a celebrated profeſſor of anatomy at Upſal; and have given the part of our authour's letter that relates to this ſubject, in his own words.

We cannot avoid being ſurpriſed at the little value our authour places on his own diſcoveries, whilſt any other anatomiſt, however celebrated, would have haſtened to publiſh them. At the end of eighteen years, he ſcarcely permits this new canal he has diſcovered, to be announced in a few lines in one of his works; at the ſame time that it has been demonſtrated at Vienna for upwards of ten years, in the common courſes of anatomy, to the profeſſors of which it was probably communicated by M. Brambilla, ſurgeon to the Emperor, and director of the military hoſpitals. Our authour ſhowed this canal to M. Brambilla, when he accompanied his Imperial Majeſty, in his travels in Italy.

Although our authour has never publiſhed this ſo long diſcovered canal, he has, however, ſhown it to a great number of his friends, and to ſeveral other perſons. — M. Murray, the profeſſor of anatomy at Upſal, in a letter he addreſſed to our authour ſome time ago, informs

forms

forms him, that the defcription of his new canal of the eye has been inferted in the laft volume of the Acts of the Academy at Upfal. *In ultimo tomo* (he writes) *defcriptio canalis a te detecti extat.* — This canal was fhown by our authour to the Swedifh profeffor, when this laft, on his coming to Italy, made fome ftay at Florence. At his return to Sweden, he addreffed our authour at Paris, where he then was, to procure the drawing and defcription of the canal, which he was defirous of publifhing in the Acts of the Swedifh Academy, for the advancement of anatomy, and for the good of his countrymen. The drawing and defcription were fent from Paris, but were loft on the road. Our authour fent him a frefh copy of them from London, but I do not know whether they ever reached him. I have inferted the figures and defcription at the end of this work, with the copy of the letter that accompanied them.

Dr. Troja, a diftinguifhed profeffor at Naples, and member of the royal academy of the fame city, in a differtation he publifhed in the year 1780 on the difeafes of the eyes, fpeaks of this new canal of the eye, obferving that the difcovery of it is due to our authour,

who

who fhowed it him at Paris in the eye of an ox.

It only depended on our authour to give this work a more original and novel air, and even to make it appear in certain refpects more perfect. He had merely to conceal the fteps by which he arrived at the truths he has difcovered, and to be filent as to the methods and proceffes that conducted him to them. The enlightened reader will find, that in pro- portion as he advances in the perufal of this work, and meets with unforefeen difficulties, the experiments our authour has imagined to furmount them, will prefent themfelves fo naturally, that he will almoft imperceptibly be led to believe, that he imagined them himfelf; fo fimple do they appear, and fo ap- pofitely are they placed. In the fame way, the new lights that are interfperfed, and the many refearches that are defcribed, feem to grow out of the matter itfelf, and not to fpring from the authour. Again, he might not have fpoken of whatever remains doubtful or undecided; he might have paffed over in filence, the enquiries he has not been able, even after fo many experiments, to decide upon. His work might have been found more complete,

becaufe

becaufe ignorance only fuffers in proportion as it is known; but our authour has throughout preferred clearnefs and utility to vain glorioufnefs and a falfe pride.

There is a clafs of readers to which this work will certainly be unpleafing, and this clafs is neither the leaft numerous, nor that which has the feweft fectators and partizans. It is compofed of thofe pretended naturalifts who explain nature at their fcrutoire; who meditating on facts feen in a wrong point of view, and copied into books, immediately divine the fprings of them; and who miftake for real caufes, the ideal ones they apply to the explanation of effects which have only exifted in their own imagination : in fhort, who prefer romances to facts, and to truth.

To folks of this defcription, accuftomed either to read, or to make, romances in phyficks, the prefent work muft appear barren, tedious, and little philofophical; and I therefore cannot exhort them to perufe it;—but to thofe, on the contrary, who are fond of facts and certain obfervations, it will be infinitely fatisfactory. I do not for my own part know of any fubject, either in phyficks or in medicine, that has been difcuffed with a greater

abundance

abundance and richnefs of experiments than the one before us.

When a work is founded on certain and new facts, we always gain by reading it, even though it is badly contrived and prefented, and contains falfe reafonings. The new truths that are found in it are real acquifitions to the philofopher, and he may eafily employ them as a bafis to truer fyftems, and to furer opinions, and laftly, to the difcovery of the true laws of nature.

But what confidence ought not an authour to infpire us with, who, after having faid, *I have made more than* 6000 *.experiments* ; *I have had more than* 4000 *animals bit* ; *I have employed upwards of* 3000 *vipers* ; finds no difficulty in adding, *I may have been miftaken, and it is almoft impoffible that I have not been miftaken !* — What a difference betwixt this authour and many others ! betwixt opinion and certainty ! betwixt ignorance and knowledge !

This work, fo enriched by the immenfe number of new facts, and by the length and difficulty of the refearches it contains, could not have been executed without the protection and conftant favours of the auguft Mæcenas

the

the authour has the happinefs to ferve. But whilft the bounties of a philofophick fovereign procure to the enlightened world fo many experiments and difcoveries, the ufe our authour has made of the means that prefented themfelves to him on his journies, will, without doubt, entitle him to the gratitude and admiration of men of letters; and it will never ceafe to be a matter of furprife, that a work which has coft fo much labour, had its birth at Paris and at London, through which places our authour, if I may fo exprefs myfelf, did nothing more than pafs.

CONTENTS

C O N T E N T S

OF THE

FIRST VOLUME.

PART THE FIRST.

b   CHAP.

PART THE SECOND.

I   PART

PART THE THIRD.

PHILO-

PHILOSOPHICAL RESEARCHES

INTO THE

VENOM OF THE VIPER.

PART I.

INTRODUCTION.

In which is fhown how little Authours agree among themfelves on the Venom of the Viper.

IT is agreed at prefent that there is no other guide in a fearch into natural truths, than a knowledge of facts ; it is only on facts that the philofopher can hope either to eftablifh a reafonable fyftem, or to form a found judgement of thofe already eftablifhed. Obfervation is alone capable of diffipating the mifts that envelop the hidden caufes of the phenomena

of nature.   And laftly, it is to the labours of obfervers that we owe the rapid progrefs philofophy has
made in our time.   But nothing retards this progrefs more, than the little agreement found amongft
authours, even in matters of experiment ; that is to
fay, in things that are expofed to the finger and to
the eye.   Nothing is more common than to fee obfervations of this kind, that are neverthelefs made
by men full of candour, frequently belied by others,
or in contradiction with themfelves.   What then is
the caufe and fource of thefe errours? Is it the fpirit of party, or is it the difficulty of nice obfervation ? Be it what it will, it is not lefs true, that
after having confulted the moft celebrated authours
on any particular we wifh to clear up, we often
find ourfelves as little informed, and in as great a
ftate of uncertainty, as we were before.   In fuch
cafes then, I have apprehended, that without being
wanting in the refpect due to the authority of thefe
great men, I might folely truft to my own eyes ;
and to render my inveftigations more decifive, I
have endeavoured by a fingular application, to difcriminate nicely, to compare the experiments of my
predeceffors with my own, to trace and develope all
the circumftances of them, and in fhort, to difcover
what may have occafioned fo great a variety in the
opinions of thefe obfervers, and in their manner of
feeing.

Such is the true motive that has induced me to
give an account of the experiments which follow.
Without this I fhould willingly have paffed them

over

over in filence, that I might not fatigue the reader, by prefenting him with what others may have already publifhed.

The enfuing experiments relate to the viper, and dwell much lefs on the anatomy and particular ftructure of fome of its parts, that on the nature of the venom of that animal. The facility with which vipers are procured at Pifa, where I made my experiments, has enabled me to multiply my refearches, and to vary them exceedingly. It would be lofing time to have no other object in view than that of rooting out the popular prejudices on this fubject that were fo very prevalent in the time of Redi. We are indebted to this authour both for his having made them known, and having difencumbered natural hiftory of them. He himfelf knew the value of time, fince at the conclufion of his letter to Magalotti he fays, *che il perder tempo a chi più fa, più fpiace. That the more any one is inftructed, the more he regrets loft time.*

When I obferved that the frequently repeated obfervations of fo celebrated a writer as Mead, clafhed entirely with thofe of Redi, I muft confefs that a glimpfe of the utility of making known the fource of the errours of thefe two great men, and the pleafure of difcovering new truths, encouraged me exceedingly to this undertaking, notwithftanding the rifk that attends the handling fuch dangerous animals.

I have deemed it neceffary to begin by making a few remarks on the teeth of the viper, and on fome

B 2　　　　　　other

other of its parts, and if in doing this I repeat certain
truths that other obfervers have already publifhed,
it is only to give a greater perfpicuity to my work,
and the impartial reader, particularly when he fees
that thefe truths are better eftablifhed, and that
the experiments which ferve as a bafis to them
have been varied in fo many ways, will readily
pardon me.

C H A P.

CHAPTER I.

Number, Structure, and Use, of the Teeth of the Viper (a).

A GREAT deal has already been written on the large or canine teeth of the viper; they had been examined, even with the microfcope, before the time of Redi, and were found to be hollow and tubulated to their very points; Redi made himfelf fully certain of this with the naked eye, and found that if they were bruifed when dry, they broke into three or four pieces, from the bafis to the point, and plainly fhowed their internal cavity; but he flatly denies that this cavity is a conduit for the yellow liquor, and that when the viper bites, this venom gufhes from the fmall hole at the extremity of the tooth. He fays that he has opened the mouths of vipers, and has always found that this yellow liquor, when they bite, runs along the outer

(a) *Note of the French Editor,* That the parts defcribed in this chapter may be more readily underftood, we have borrowed feveral figures of the head of the viper from Mead's works: fee Table I, and the explanation that precedes it. We requeft the reader to likewife give a glance at Table II, before he goes any further.

part

part of the tooth from the top to the bottom, and that it never flows from within. *I have assured my-self well of this, continues Redi, by several experiments, and by the often repeated testimony of my own eyes.*

The celebrated Valisnieri adds moreover, that the canine teeth of the viper are pierced with four very small lateral holes, through which he believes that the most subtile part of the venom penetrates into the wound from within the tooth, whilst the thicker and grosser part of it runs along the outer surface. Mead and Nicholls, on the contrary, set out from the analogy betwixt the viper and the rattle-snake, in the last of which this humour is very dif-tinctly seen to flow from the inner part of the tooth, and maintain that the venom of the viper flows likewise from the point of its canine teeth, or at least from an opening they have towards their extremity. I have several times repeated Redi's experiments, by opening the mouths of these animals when living, and acknowledge that I have never been able to assure myself whether this venomous liquor issued precisely from within the tooth, or whether it simply glided along the outer part, from the basis to the point. If I held the head of the viper with the point of the teeth downwards, I had only to press strongly against the muscles of the palate, to make this yellow liquor flow rapidly from the basis to the point of the tooth ; but if I held it with the teeth turned upwards, I saw the poison collect in-stantly about the basis of the tooth, and entirely fill the sheath or bag that serves to enclose it. Redi main-

tains

tains moreover, that this sheath is the true reservoir in which this humour is deposited and preserved, and thinks it is secreted by a small neighbouring gland situated below the orbit. Nicholls says, on the contrary, that there is a vesicle, or small bag, separated from the sheath, and that this gland is destined to quite another purpose, as the secreting some lymphatick or salivary humour.

In this uncertainty, I conceived that the best step would be to examine by my own proper observation, the structure of the viper's teeth, and so inform myself well of their uses, since the descriptions these authours give of them are obscure, and the observations of Mead and Nicholls contradictory to those of Redi.

At each side of the anteriour and upper part of the head, the viper has a moveable bone which forms in part the upper jaw; each of these two bones has two sockets at the side of each other, which are only separated by an immoveable, but very brittle lamen, the substance of which is spongy and like that of the bones themselves: in these sockets are fixed the canine teeth, which are sometimes found to the number of four, seldom of three, and oftener again of two. It is observed that when these teeth are four in number, they are not all fixed in their sockets with the same strength and stability: there are at that time usually two, or one at least, moveable, and capable of being easily pulled out without breaking: this cannot be done with the others, which are never pulled out entire, notwithstanding they

have

have no roots like ours. I have sometimes found three moveable ones. I have likewise seen some vipers that had only two canine teeth, both of which were however weak and moveable; but this is a very rare case.

At the basis of these large teeth, and quite out of the sockets, six or seven very small teeth are invariably found; they sometimes even amount to the number of eight. When they are examined attentively with a magnifying glass, they are found to be fastened at their basis by a kind of web of a fine and soft membranous texture. These small teeth diminish in size, in proportion as they are more distant from the sockets of the canine teeth; those which are nearest the sockets are likewise the hardest and best formed. The other smaller ones are more tender, more imperfect, and as it were mucous, particularly at their basis; they seem in reality to owe their formation to a whitish and gelatinous matter.

Besides the two kinds of teeth I have mentioned, the viper has still another order of them, much more minute than the former, and resembling small hooks; they are strongly fixed to the number of ten, eleven, and sometimes fifteen, in two small, pretty long, parallel, bones, which on each side form the upper jaw; and of eight, nine, and sometimes twelve, in each of two bones that form the lower jaw.

The canine or large teeth, as well as the other smaller ones found at their basis, are enclosed in a bag or sheath, which covers them on all sides, and is com-

posed

pofed of very ftrong fibres, and of a cellular web.
It is however open towards the point of the tooth,
and terminates there by folding together its two
lamina, which at their junction are often dentated.
This fheath feems to be a prolongation of the external membrane of the palate.

The canine tooth is feldom more than three lines
in length, Paris meafure (a). Its bafis is not more
than half a line in diameter; its figure is that of a horn
a little flattened, and fomewhat bent towards its
bafis.

This tooth terminates in a very fharp point, towards which it infenfibly lofes its curvature, and becomes at length almoft ftraight. Below the middle
of the tooth, towards its point, and in the convex
part, a fmall opening is vifible to the naked eye, very narrow but exceedingly long, which ending in a
fmall, channelled flope, that can fcarcely be feen
with a microfcope, terminates in this way at the
point. Hairs plucked from the whifkers of foxes,
cats, dogs, &c. may be readily introduced into this
opening, and it appears, when viewed with a microfcope, to be a cleft, the length of which is almoft
a fourth part of that of the tooth, and the breadth
the fixteenth part at moft. It reprefents with its exteriour edge a very long or flattened ellipfis, becoming larger towards the bafis of the tooth. This flope

(a) The French, or Paris inch, is fomewhat larger than the
Englifh one, but not fo much fo as to make a fenfible difference
in the line, which is its twelfth part---Therefore whenever the
latter is introduced in this work, it may be confidered as the
twelfth part of an Englifh inch.

penetrates interiourly, and is terminated at its two
fides by two fhort edges or lips, thick and raifed up.
Another opening is likewife found in the convex
part of the tooth, towards the bafis, and near the place
where it is fixed in the focket.   This opening likewife begins by a fhallow cavity, immediately at
leaving the focket.   It is much larger than the firft,
but not longer.   In proportion as this flope or fmall
hollow penetrates into the tooth, it pierces it for its
whole length, and forms a channel which terminates
in an elliptical hole at the point.   A bit of filk is
eafily paffed from one opening to the other, particularly when care is taken to introduce it at the bafis,
where the natural entry of this paffage is found. The
fide of the fecond opening refembles a parabole, the
bafis of which paffes over the bony edges of the focket, and the other fides of which end in a fomewhat
obtufe point towards the fummit of the tooth.   The
canine tooth of a viper then is hollow and tubulated
for its whole length, from the bafis to the point,
and has two holes in its convex part.   This hollow
is however not fuch as one would fuppofe it to be,
on viewing the third figure of Mead, and the def-
criptions of Redi.

   The tooth of the viper has a double pipe or tubule
almoft for its whole length, a circumftance hitherto
entirely unknown to obfervers. Thefe two canals
or tubes do not communicate with each other, and
are feparated by a bony partition, very brittle towards
the bafis, but which becomes fomewhat ftronger in
proportion as it advances towards the point. One

4   of

of thefe tubes or canals, which I call the external
one, becaufe it is at the fide of the convex part of the
tooth, begins, as has been feen, at the bafis of the
triangular opening, and goes on enlarging by degrees to the middle of the length of the tooth,
whence it gradually narrows, and ends at the elliptical opening of the point. The inner canal, on the
contrary, which is towards the concave part of the
tooth, begins with a large opening at the bafis,
from whence it advances, clofing by degrees, and terminates at length in a blind point above the middle
of the tooth. The partition likewife that feparates
thefe two cavities is crooked, and its convex part
is turned towards the hollow of the canal it terminates, fo that its bony fubftance rather prefents an
irregular, curvilinear figure, or truncated cone,
than a perfect cone itfelf. The blind canal communicates with the focket in which the tooth is
fixed, and receives veffels and nerves, which enter
by a fmall oval hole, perceptible to the eye, and
opening at the edges of the focket itfelf, towards
the inner part of the jaw. This bone of the jaw has
likewife a large round opening, which begins a canal
placed a little below and at the fide, and opening
one way into the focket, and the other laterally and
more below, towards the furface of the fame part
of the jaw.

The fmall teeth placed at the bafis of the large
ones, refemble them very much both in their inner
and outer ftructure. Thofe particularly that are
placed the neareft the firft, and which are the firmeft,

are

are perfectly like them, unless it be that their basis is not so well finished. Like the large ones, they have all the elliptical hole towards the point, and a part of the triangular hole at the basis. The two conduits, internal and external, are also seen in them.

It is not the same with the other very small teeth I have mentioned, which are far the most numerous, and in both jaws. These are not channelled, and have no kind of opening, either at the point or basis.

C H A P T E R   II.

Of the yellow Liquor that flows from the Tooth of the Viper.

WHEN the viper wishes to bite, its canine teeth are raised by a mechanism, which Nicholls has described perfectly well in the anatomical appendix he has annexed to Mead's work on poisons. But those of the large teeth that are not so well fastened in their sockets, are less capable of raising themselves, in proportion as they are more moveable, and badly fixed in the jaw. Nicholls pretends, that when one or two of these four large teeth are moveable, the viper can only bite with one tooth on each side. Indeed he does not found his opinion on any experiment,

riment, but feems to account for this by certain final
caufes which I cannot admit, fince in phyficks thefe
kind of proofs become of no weight.   He remarks
that there is fuch a diftance betwixt the two canine
teeth of the rattlefnake, that the yellow fluid, which
is carried by a conduit betwixt them, would entirely
enter into the fheath, inftead of being conveyed to
the animal bit by this fnake : on this account he
does not hefitate to believe that the conduit of this
liquor is precifely fixed oppofite the hole at the
bafis of the fingle tooth on each fide, with which
the viper feizes what it bites.   But befides that no
organs are feen to execute this function, and that the
mechanifm of it cannot be inveftigated, I can take
upon me to fay that I have fometimes feen all the
four canine teeth of a viper equally firm and ftrongly
fixed in their fockets, and have ftill more frequently
found three of them, well fixed and very ftrong, in
a ftate both to feize and bite. It is not to be doubted
but that the viper, inftead of fimply biting with two
teeth, one at each fide, muft feize equally with all
thofe that are firmly fixed in the fockets, and I have
even affured myfelf of this by experiment.   It is
not true then, as Nicholls pretends, that the conduit
of this yellow liquor only adapts itfelf to one fingle
tooth on each fide when the viper bites ; befides, this
fpace which is obferved betwixt the canine teeth of
the rattlefnake, is not alike found in our vipers, the
teeth of which, almoft from the bafis to the point,
touch and embrace each other in fuch a way, that
no fluid can pafs betwixt them, much lefs this yel-

3   low

low venomous liquor which is somewhat glutinous. It is moreover determined that the viper bites and seizes not only with the teeth that are fixed in their sockets, but likewise very often with those that are moveable. Of ten vipers I examined, there were three that had two moveable teeth, and two firm in their sockets; the seven others had only one moveable tooth, and two firm and well fastened. If I except one of the first three vipers, and two of the seven last, all the others to which I held a bit of tendon of beef, boiled and well stripped of its coat, seized it forcibly, and left the marks of their teeth strongly printed in it: I must however observe that their least firm teeth were not of the most moveable kind, and that when they are very loose I have found them to rise so little that it is absolutely impossible for their points to touch the body seized by the viper.

Since it is certain that this creature never bites without a risk of losing some of its teeth, Nicholls conjectures with great sagacity, after Redi, that nature has intended the small ones at the basis of the others, to replace, when there is a necessity for it, those that the viper loses from time to time. Their crooked shape renders it difficult for them to be drawn from the wound, and in the course of my researches I have sometimes observed, that not only those which are moveable, but even the firmest of them, are alike subject to accident. The thinness of the tooth, and the strength of the animal that has been bitten, contribute equally to this loss; and this opinion becomes still more probable when we consider that these small moveable teeth are ex-

actly

actly formed like the canine ones, that is to say, that they have likewise two canals, (tho'e however that are the moſt perfect) and the ſame openings at their baſis and at their point. But all theſe reſemblances were in ſhort but one reaſon more why experiment ſhould be confulted, and the truth eſtabliſhed by nice obſervations.

I have ſometimes obſerved in one of theſockets, a very moveable tooth, the ill-formed and ſtill gelatinous baſis of which faſtens itſelf to the edges or ſides of the hollow ; this tooth may even be drawn a little way out of the ſocket without detaching it entirely, by means of a tender and mucous matter that ſerves as a glue to it. On moving the jaws, I made the one next it raiſe itſelf very well, but the one of which I have ſpoken, abſolutely continued reclined on the baſis of the moveable bone of the jaw. It is clear that this tooth had been of the number of thoſe that are at the baſis of the great or canine ones.

. I expreſsly drew from a large viper one of theſe laſt, which was moveable and ill fixed in its ſocket, and obſerved ſome time after that the largeſt of thoſe which are placed beneath the ſheath and beneath the ſocket, had advanced a little towards the empty ſocket. Some days after I thought I perceived it to have approached ſtill nearer. I purſued my obſervations on every ſecond day, and at length ſaw that this tooth had perfectly fixed itſelf in the ſocket, where however it was as yet very moveable, and badly faſtened. At the end of thirty days it was fixed in ſo ſolid a way as to be capable

of

of biting. The neceffity one is under of frequently handling the viper to be fatisfied of the ftate of its teeth, and of opening its fheath with pincers, or with the blunted point of a bit of wire, makes this experiment very dangerous. The repeated compreffions the fmall teeth receive, from the contraction of the mufcles of the jaw, and the action of the fheath itfelf which conftantly preffes on the points of thofe teeth that are moft raifed, are quite fufficient to pufh the root of the tooth in queftion, into the focket which the falling out of the old tooth has left empty.

The laft or fmalleft teeth are certainly not employed in biting, but are intended to draw nearer to the throat, and to hold firmer, the animal the viper has already feized.

The fingular ftructure of the canine teeth, fo different from that of the other teeth of the two jaws, is a powerful perfuafive of its being from them that the yellow liquor flows ; it is not however without fome apparent reafon that Redi, otherwife fo exact, has been led into an errour.

To fully affure myfelf of this circumftance, I bound the head of a viper I had juft killed, to a table. To diftinguifh better and to a greater certainty, I took the precaution to remove the lower jaw : the canine tooth, in the way I had fixed the head, was turned upwards, and I obferved the elliptical cleft with the ftrongeft lens of Ellis's microfcope. I gently compreffed the palate with a fomewhat blunted iron, when a flightly tranfparent yel

low

low liquor, which formed itſelf into a drop, and at
length glided along the outer ſurface of the tooth,
inſtantly appeared at the elliptical hole of the point.
I repeated this experiment ſeveral times, and always with the ſame ſuccefs. I afterwards cloſed the
ſmall opening with wax, and then compreſſed the
palate, but not a particle of the venom ſhowed itſelf.
I however ſaw it through the tranſparent ſides of the
tooth, conveying itſelf from the baſis to the point by
the external canal, which it had entirely filled. I
put a round bit of wax with riſing edges all about
the tooth, in ſome other heads of vipers, immediately below the elliptical hole, and having ſtrong-
ly compreſſed the palate, I inſtantly ſaw the yellow
liquor guſhing forcibly, and as it were by ſtarts,
from the point, and ſcattering itſelf abundantly on
the piece of wax, which it entirely covered all
round the tooth.

I have likewiſe been able, although with difficulty, to cloſe the hole at the baſis with wax, and I
have then found it to be in vain that I compreſſed all
the muſcles of the head ſucceſſively. I could never
force a drop of venom from the point, nor could I even
diſcover it through the ſides of the tooth. Whenever a viper is held in the hand with the teeth turned upwards, it is eaſy for an attentive eye, accuſtomed to obſervation, to ſee this drop of yellow liquor preſent itſelf at the elliptical opening, in ſuch
a way that it may be more or leſs encreaſed at will.
I have repeated this experiment a thouſand times,
and have invariably ſeen the ſmall drop of venom

ex-

exuding from the elliptical hole of the tooth. Nay, what is more, on compreffing violently, this liquor is fometimes obferved to force itfelf out fuddenly, and to fpirt to a confiderable diftance. It muft however be remarked, that when the tooth is once wetted with it, particularly when it is entirely covered with the fheath, this humour, or the drop it forms, glides fo very precipitately along the tooth, that it is fuddenly feen at the bafis without having been perceived at the point. It in this way imperceptibly fills the fheath, fo that it is difficult to perfuade ones-felf that it really iffued from the point of the tooth. This is the way that fo exact an obferver as Redi has been led into an errour. It is not proper, after his example, to employ living vipers, nor to open their mouths forcibly, fince the flowing out of the venom will then be too fudden, and fince it will then be dangerous to obferve as nearly as is neceffary to prevent being deceived.

I not only faw the yellow liquor flow from the point of the tooth that I particularly obferved, but likewife from the neighbouring tooth, when there was one, fo that it proceeds equally from all the canine teeth at a time, not excepting thofe which, without being altogether firm in their fockets, are however fufficiently fo to rife with the others. In a word, in all the heads of vipers I have obferved, I have feen this humour conftantly flow from all the canine teeth that raife themfelves fufficiently, on the compreffion of the mufcles of the palate, and on opening the mouth with a force that would be capable
ble

ble of wounding an animal feized by the viper. This fhows that Nicholls is deceived, when he pretends that the venom only flows from one tooth at a time on each fide.

CHAPTER III.

Of the Part where the Refervoir of this yellow Humour is feated.

THE yellow liquor iffues then from the point of the vipers tooth, contrary to the fentiments of Redi, who regarded the bag or fheath that envelops not only the canine teeth, but likewife the others that are found at their bafis, as the true refervoir of this venom. This opinion is ftill again belied by the ftructure itfelf of this fheath, which has a large aperture next the cheeks, through which this liquor would inceffantly flow with the greateft eafe ; fo that every time the jaws of the viper fhould be extended, the venom would be conftantly feen oozing through the extremity of the fheath, even though the viper fhould not bite. This is what no one has hitherto obferved. It is befides certain, that when this fheath is opened with fciffars, neither this yellow humour, nor any kind of fluid that may have collected there, is found in its cavity.

But fince, as has been feen above, this liquor flows from out the elliptical hole at the point of the tooth, it muſt neceſſarily be carried to the aperture at the baſis by another conduit than this ſheath, in which it is certain that no remains of venom are found. It will not be difficult, according to this hypotheſis, to diſcover the ſmall veſicle that is really deſtined to contain it.

If after having ſtripped the teeth of this bag or ſheath, a preſſure is made on the palate, this humour is obſerved to flow through a ſmall and almoſt imperceptible hole, placed at the anteriour part of the maxillary bone, within the ſheath, and at the ſide of the baſis of the canine teeth ; ſo that when theſe teeth are covered by the ſheath, this ſmall orifice comes in contact with the inferiour opening of the tooth. Indeed with the help of a magnifying glaſs, a very ſmall orifice is diſcovered, ſituated in the midſt of a ſmall cleft or furrow, which anſwers to the maxillary bone. I endeavoured to introduce into this orifice a fox's hair, very fine, but nevertheleſs pretty ſtrong, and at length ſucceeded in paſſing it quite acroſs the ſheath, by a long membranous conduit, into a ſmall veſicle placed beneath the muſcles of the upper jaw, on its lateral part. It is a membranous bag of a very ſtrong and cloſe texture, which is again partly covered by tendinous fibres. Its ſhape is nearly that of an equilateral triangle. It differs from other veſicles, which are crooked or ſpheroidal, inſtead of which the baſis of this is in a manner ſtraight. This ſmall veſicle terminates next the eye in a tranſparent canal, which after having proceeded beneath the orbit for the

ſpace

space of two lines, pierces the sheath, and at length opens at the extremity of the sockets, into the small cleft of which I have spoken. When this conduit reaches the vicinity of the sheath, it dilates a little, and it is here that the venom finds the greatest obstacle to its passage, which is owing to the compression it meets with from the bones of the jaw.

The vesicle I have spoken of, and which serves as a reservoir for this humour, is three or four lines in length, and two at most in breadth at its basis. It never contains more than four or five drops of the venom, which is forced from it principally by the action of a strong and powerful muscle, that rising out of the lower jaw, folds inwards a little, then makes an arch, and preceeding to the upper jaw, runs over a part of it, and fastens itself there. Towards the inner angle of this constrictor muscle, or rather towards the part of its curvature nearest the upper jaw, the small vesicle begins ; it is covered with this muscle for almost the whole of its length. Thus placed, it is as it were enclosed in a press; and is bound and fixed to the adjacent bony parts, by means of two tendons, and of the canal, so that it can neither force itself forward, backward, nor sideways, and must necessarily bear the double action of this muscle, which now compresses it, when the viper bites and presses forcibly, and now again suffers it to dilate, when the muscle itself contracts and enlarges. That which proves this muscle to be chiefly intended to force the venom from its reservoir is, that it is fastened to each jaw in such a way as to be

of

of very little ufe to the creature in clofing its mouth, which clearly cannot be its principal purpofe.

The hairs of a fox's whifkers penetrate and pafs eafily from the veficle through the excretory duct, and go out at the orifice fituated at the inner part of the fheath. I have fometimes fucceeded in bringing them even to the elliptical hole at the point of the tooth. This is certainly the route the venom takes for the purpofe of going out at the fmall orifice of the fheath, which correfponds precifely with the height of the parabolical hole of the tooth *(a)*.

As the fheath is very nicely fitted to the bafis of the canine tooth, the venom that goes out of the conduit at the fmall orifice, muft of neceffity enter entirely into the hole of the tooth; and although it gufhes in abundance through this canal, it does not at all fcatter itfelf in the fheath, fince the orifice out of which it flows is infinitely fmaller than the parabolical hole to which the nice fitting of the fheath makes it immediately correfpond. In a word, it paffes entirely into it, particularly when there is only one of thefe teeth. Still more, I have obferved on folding the fheath back above the bafis of the teeth,

(a) It muft appear very ftrange that Doctor James, who has written after Mead, has afferted in his Difpenfatory, that the true refervoir of this liquor is the bag (fheath) which covers the roots of the large teeth of the viper, and that a fmall veficle is found at the top of this bag, which opens at its extremity, to give a paffage to the teeth that fhed the venom. It appears however that this writer made many experiments on the viper, and with the intention of making them well.

and

and preffing flightly and gradually upon the conduit, that the venom is carried by a natural declivity towards the hole of the tooth, which it entirely
fills before a drop of it is fpilt in the fheath. This
natural declivity is fimply caufed by a fmall hollow
in the jaw, which extends as far as the parabolical
hole, and is fcarcely feen with a microfcope. I do
not however deny but that in fome particular cafes
this liquor may flow directly into the fheath, and
even glide from thence to the points of the teeth,
particularly when two of them are fo clofe to each
other as to touch and leave nothing but a hollow
betwixt them, and when the viper bites in fuch a
way as to force its teeth deep into the fiefh, and to
ftop up the parabolical hole; it muft here fqueeze
with fufficient force and long enough to comprefs
the veficle, and give time to the liquor to glide betwixt the two teeth. In thefe cafes, which cannot
however but be very rare, there is no doubt but that
the animal may even kill without the venom having
made its paffage through the ufual conduit of the
tooth. I repeatedly ftopped up with pitch, fometimes the parabolical hole, at others the elliptical
hole, and fometimes again both of them, and found
that the yellow liquor did not then reach the bottom of the fheath but with great difficulty, and after a ftrong compreffion had been made for a long
time on the conftrictor mufcle. I lay it down from
hence, as an infallible conclufion, that the venom
flows from the point of the tooth, and not form the

bag

bag or sheath, whether the viper of itself causes it to flow in biting, or whether a compression is purposely made on the vesicle I have spoken of.

C H A P T E R   IV.

The Venom of the Viper is no other than the yellow Humour that flows from the Tooth when the Viper bites.

It happens very often, in vipers particularly that have been lately killed, that this yellow humour dries, stops up both holes, and totally obstructs the canal of the tooth.   As it cannot then enter into the tooth so as to find a passage through it, it must consequently flow from the excretory conduit into the sheath.   This observation is so much the more necessary, as without it 'twould be easy to fall into an errour, and to presume that the venom, instead of being conveyed by the tooth, flows from the sheath, and is carried from thence into the wound.

I was desirous of assuring myself how far one may rely on the opinion of those who believe that the bite of the viper is only mortal on account of the rage the creature is thrown into before it bites. I omit the mention of an infinite number of experiments I have made to be certain, with Redi, that the yellow humour which flows from the tooth of the viper is mortal when introduced immediately into the blood h e mediu m of a wound.   I shall only observe, that the various experiments of Redi and Mead

agree

agree perfectly as to the truth of this circumstance, and I cannot conceive how certain celebrated writers have continued to perfuade themfelves to the contrary, and to attribute the mortal effect of the bite of the viper to the rage of the animal, and to the power of the exalted ftate of the faliva, rather than to the fpecifick character of this humour.

I have frequently enraged vipers, and afterwards opened the mouth in fuch a way that they could neither comprefs nor bite with it. I have then foaked bits of cotton well in the foam or faliva with which it was covered, and applied them to the wounds of animals, the bleeding of which had ceafed. I have never feen any accident caufed by this, nor have the animals appeared to be the leaft difordered. Neither the foam then, nor the other humours of the viper's mouth, are capable of caufing death when introduced into the blood of an animal.

I have fevered at one ftroke the heads of feveral vipers from their bodies, at a time when far from being enraged, they were in a calm and tranquil ftate. I have then taken the venom from the tooth itfelf, to be fure of having it pure and unmixed. I have taken it from fome of them immediately on feparating the head, and from others fome hours after, when the head had dried in a great meafure, and had ceafed to move. On applying this venom carefully to the wounds of different animals, they have all without any exception been killed by it. We muft conclude then that the humour alone which flows from the tooth, has a deadly quality,

to

to which the rage of the animal does not at all contribute. But to obviate all objection, and to prevent the being reproached with having neglected
to make a viper bite after having enraged it, and
having contented myself with introducing its foam
into wounds, I took one, and made it bite several
animals. When I conceived that there could be no
longer any remains of the venom, I began to prick
and torment the creature, and in a word, employed
all the means I could think of to enrage it. When
I saw by its hissings, and the rapid vibrations of its
tongue, that it was become furious, I presented animals to it, which it bit with all the force it was capable of. Neither of them however died, or seemed
to feel any inconvenience. This was a very natural
result, since the liquor that flows from the tooth,
which alone has the faculty of killing, had already
been entirely wasted, and since nothing more now
remained than the foam and other humours that are
in no way venomous, even during the most excessive
rage of the animal. I repeated this experiment on
two other vipers, with the very same success.

I was desirous of making another experiment,
which, to prevent its being dangerous, requires a
great deal of precaution and address on the part of
the observer, although after all it cannot be more
decisive than the preceding one. This was to entirely remove the two vesicles that contain the venom. After several fruitless attempts I at length
succeeded, without doing much injury to the viper,
and without tearing its mouth. I made an incision

into

into the ſkin that covers the two veſicles, and having ſeized them with pincers, cut them entirely
out with a biſtoury, Thoſe who are accuſtomed to
diſſect this ſpecies of animals, muſt ſee clearly that
this experiment is attended with more danger than
difficulty. To ſucceed in it, the neck of the viper
muſt be ſeized by ſome one, or it muſt be well tied
to a table, in ſuch a way that the creature cannot
raiſe its head to bite. Having removed the veſicles,
I had two frogs bits, ſo as to diſcharge whatever venom might remain in the teeth or in the extremity
of the conduit : neither of them died. I kept this
viper a long time, and at different times made it
bite both large and ſmall animals, as well with warm
blood as with cold. Neither of them died, nor had
any other complaint than what muſt have neceſſarily
been cauſed by the ſimple mechanical wound of
the tooth.

CHAPTER V.

*The Venom of the Viper is not a Poiſon to the
Viper itſelf.*

Very grave authours have likewiſe imagined, that
this humour, which deſtroys other animals, is not
leſs hurtful to the viper itſelf ; and this is the opinion of thoſe who have written lately on the venom
of animals. The examples of ſcorpions and ſpiders

4   which

which kill each other with the bite or fting, feemed to favour this opinion very ftrongly. We read in the philofophical tranfactions that rattle-fnakes die in a very few minutes, when they bite each other. It is at prefent known that this fnake is a fpecies of viper, larger than ours, and therefore they have by analogy drawn the fame conclufion as to the viper and other venomous animals.

Certain Spaniards had brought from the Eaft Indies three fnakes called *Cobras de capello,* and one only had furvived the frequent combats they had amongft themfelves. Doctor Mead concluded that the other two died of the venom, and confequently that the viper's venom ought to be likewife mortal to the viper itfelf. It feems to me that he ought rather to have drawn a contrary conclufion, for it is not likely but that the victorious fnake which furvived, would have been fometimes bit by the two others; and yet we fee that it contrived to live.

'Twould have been undoubtedly much better to have made experiments, than to have founded an opinion fo flightly on a mere matter of fact as Mead has done, and on a fimple analogy drawn from cafes that are very rare; particularly as the fury with which fcorpions and fpiders combat and mangle each other does not prove that they die of the venom they have received. Befides, it has been obferved that the fpider which leaves the combat victorious, only dies when it has loft fome one of its organs neceffary to life. As to the rattle-fnakes, the examples we have of their combats are too rare and not fufficiently authenticated

ticated to furnifh a good analogy. This could befides be never any thing more than a fimple analogy, rendered fo much the weaker by there being
certainly a very great difference betwixt this fnake
and our viper, whether we regard their ftructure,
or the activity of their venom.

It is not eafy to provoke vipers to bite each other,
whatever care may be previoufly taken to kindle
them to fury. I employed the following method to
get the better of this repugnance. Having feized
the neck of a viper with pincers, I kept its tail in my
other hand to manage it with greater fafety. I employed an affiftant to feize a fecond one in the fame
way. I held the body of one of them to the head of
the other, which perceiving itfelf to be clofe grafped
by the neck, hiffed, twifted itfelf, and fell with
fury on every thing that came near it. The former
one, which it bit feveral times, was much fmaller,
and expreffed each time, by the livelinefs of its motions, the violence of what it fuffered. I found a fuperficial wound at the part where it had been bit, moiftened with blood and with the venom from the tooth.
I enclofed it in a glafs vafe, where it continued tranquil for fome minutes. Two hours after I found the
part where it had been bit a little fwelled; this
fwelling however continued but a little time, and
the viper, recovering its natural vivacity, crept
along the fides of the vafe, and raifed its head with
the fame ftrength as if it had not been bitten. After twelve hours I fet it free, when it appeared as
strong

ftrong as another one I placed with it by way of comparifon. I put it again into the vafe, and on the following day found it ftill as lively and healthy as before. At length, thirty-fix hours after, feeing no appearance of its having been envenomed, I killed it. I found feveral holes in the fkin at the part where it had been bit; the mufcles themfelves of the back were very deeply pierced ; and the blows of the tooth had in more than one place forced it through the body, and through the abdominal vifcera. And laftly, the wounds were a little inflamed, but there was no appearance of fwelling or tumour.

Two days after I tried two very large vipers, which threw themfelves furioufly on the animals that were prefented to them. I made them bite another middle fized viper, which received, from one of them two very deep wounds made with the teeth, from the other four. One of them even left a tooth in the wound. At every blow the viper received on its belly, all of which were directed to the fame fpot, it gave violent fymptoms of pain, hiffed, and nearly efcaped from the hands of the perfon that held it. I put it into a glafs vafe, in which it remained for a few minutes in a ftate of infenfibility : however, on afterwards placing it on the ground, it crawled about with great agility. I could never difcover any fwelling at the part where it had been bit; the fkin had notwithftanding been lacerated, and the mufcles laid bare : there was no hemorrhage. I kept it four days in the vafe, during

ring which time it did not appear in the leaft difordered. The fecond day I held an animal to it, which
it inftantly bit, and which died two hours after. I
at length killed it and found that the blows of the
teeth had pierced it through and through: the
wounds were fomewhat inflamed. The fame thing
happened to five other vipers which I had bit repeatedly. I even forced a fixth to bite itfelf at the
tail. Neither of them died, nor appeared to be in
the leaft difordered.

But to prevent any one from thinking that the
hardnefs of the fkin had kept the venom from penetrating, and to introduce it with greater certainty
into the blood, I removed a confiderable portion
of the fkin from the backs of four vipers, and had
them bit by feven others, from which they actually
received feveral blows of the teeth ; neither of them
died, or became ill, and only one of them appeared to be heavy and languid, and had a fwelling at
its back.

Again, I irritated another viper, by pricking it
on the body with a pointed iron, and afterwards
made it bite a piece of jagged glafs. The venom
fpread from the tooth over the whole mouth,
which the glafs had cut and made bleed. I kept it
ftill, and waited the event. For the three firft days
it crawled about a little. On the fourth it was more
lively, but did not yet make any attempt to bite,
even when provoked. On the feventh day I opened
its mouth, and found it entirely healed, without
any fcar. On the fame day I made it bite a fmall
animal, which died an hour after.

I re-

I repeated this experiment on three other vipers, and employed the following method. From one of them I removed a portion of the ſkin of the neck, from another a portion of that of the back, and laid bare the muſcles of the third juſt above the tail. I wounded each of them at the part laid bare, bending the point of the lancet a little, to open the wounds the better. Into each of theſe wounds I introduced a ſmall drop of the venom, that is to ſay, as much as was neceſſary to entirely fill it. I afterwards returned theſe vipers, each into its vaſe, where they remained very tranquil, and ſeemed to have ſuffered but little. Their wounds were however inflamed, but there was no ſwelling. I kept them alive for ſeveral days.

We now ſee what opinion ſhould be entertained of the analogy betwixt the venom of the viper and that of other animals, and may judge into how great an errour thoſe have fallen who have believed that the yellow humour which flows from the tooth of the viper, and which is mortal to other creatures, is likewiſe ſo to itſelf, and that the bites of theſe dangerous animals are capable of poiſoning each other. If analogy could have any weight in this inſtance, I ſhould be tempted to belive that, contrary to the opinion of Mead, the venom of the ſcorpion would have no ill effect on the ſcorpion itſelf, and that there is probably no venomous animal on earth, the venom of which can be hurtful to thoſe of its own ſpecies. If it be ſo, it can only be in a very few animals, and only in the ſmalleſt of them, the venom

of

of which is acrid and cauftick, fuch as bees, wafps, and hornets. It may likewife be true that the fcorpions of Afia and Africa carry a venom that is mortal to their own fpecies, fince the venom of the Italian fcorpion, when put on the tongue, is acrid and pungent. It appears to me, that the general errour which many obfervers, who are otherwife very exact, have embraced, has its fource in a deceitful experiment. It had been remarked, that when a fcorpion was furrounded by live coals, it was firft agitated, and then turned its fting towards its back, as if to wound itfelf. As it at length died, and was even burned up, from its violent agitation amongft the live coals, it was fimply believed that it died from its venom, and from its own wounds. The experiment is equivocal; it is even falfe. I have repeated it a thoufand times, and have never obferved that the fcorpion ftruck itfelf with its fting; it died roafted and burned up, and not envenomed.

It has likewife been obferved, that the frefh water polypus, in fwallowing its prey, fometimes fwallows the arms or claws with which it holds it; and likewife, that when two of thefe polypuffes difpute together, the ftronger frequently prevails, and fwallows the claws of the weaker. The polypus, however, in neither of thefe cafes dies, although its venom, as we fhall fee hereafter, is very active. The parts thus fwallowed, leave the ftomach foon after, entire and alive, without having fuffered any apparent change, and continue as before to ferve as claws to the animal.

VOL. I. D CHAP-

CHAPTER VI.

The Venom of the Viper is not a Poison to every Species of Animals.

Thus far we have seen that the venom of a viper is neither a poison to the viper itself, nor to those of its own species ; and this singularity led me to suspect that it might also be innocent to some other kinds of animals. Indeed why should it not be so, as well as to the viper ? If, in a word, it is not capable of decomposing the solids, and altering the fluids, of any particular living machine ; if it can neither disturb the harmony of it, nor occasion death ; why should there not be other living organized creatures, on which it may ha⁻ ? as little action ? We know but little of the manner in which poisons in general act, but we know that there are many very active substances which produce the most terrible effects on certain parts, and which notwithstanding have no effect on others. Stibiated tartar, for example, a substance that is introduced without danger into the eyes, is a very violent emetick when received into the stomach. There are persons who are thrown into convulsions by the fragrant smell of the rose. These various accidents are owing without doubt to the structure and organization of the animal machine.

machine. Certain fubftances are known to be poifons to certain animals, whilft far from being hurtful to fome others, they even ferve as an aliment to
them. Such is hemlock for inftance, which deftroys
the human fpecies, and nourifhes goats. It is thus
that the bitter almonds we eat on account of their
flavour, kill certain birds, and do no injury to others.
It may likewife be the cafe then, that the venom of
the viper may not be a poifon to all kinds of animals,
particularly if it act like narcoticks, that do not
caufe death by corroding the folid parts. Corrofive
fublimate is a poifon deftructive to all living animals,
becaufe its mechanical action is in fact capable of
exercifing itfelf on all the organs of the animal
machine. Narcoticks, on the contrary, fo dangerous
to men, produce no ill effect on dogs. The different ftructure of the organs of animals may then be
the occafion, that the fame fubftance may at the fame
time be a very active poifon to certain fpecies' of
them; and altogether of an indifferent nature, or an
aliment, or even an excellent remedy, to others.

It is on thefe conjectures that I engaged in the long
courfe of experiments I am going to relate. I had
already obferved, that of all animals, leeches are in-
conteftibly killed with the greateft difficulty. When
they are cut in pieces, each portion preferves for feveral months the motions it had before it was feparated from the others. I conceived that an animal fo
tenacious of living, might well refift the venom of
the viper, without dying or even being incommoded

by it. I fixed then upon leeches, but before I had them bit, I took care on removing them from the water, to wipe them very dry with a piece of linen, fearing that the mucofity or kind of glue that covers them, and which they emit when touched, might prevent the fuccefs of my experiment. I had one of the largeft kind, that are called horfe leeches, bit by a very ftrong viper which I had previoufly thrown into a violent rage. It pierced its body, from which a fmall quantity of blood flowed, feveral times through the whole fubftance of it. I afterwards put the leech into water, and it continued to move as ufual. On the day following I changed its water, and found it very lively and fwimming perfectly well in the glafs. It lived in this way for feveral days, and would certainly have lived much longer, had I not applied it to another purpofe.

I took a fmaller one, of that fpecies that have different coloured ftripes on the back, and that are employed in medicine. I had it bit by two vipers, which pierced its body in feveral places. It was bit the next day by a third, and the day following again by two others. Its fkin was full of holes, from which a vifcous and black matter flowed, on preffing with the fingers. Notwithftanding this it continued to live, and move about in the water. Laftly, I had many other leeches of both kinds bit in the fame way, at fome times in the head, at others in the body, and neither of them died of the venom.

I did

I did not ftop here, but fearing that the venom
might have been enveloped and deadened by the
glutinous humour of the leeches, which oozes out
in greater abundance the moment the tooth of the
viper has pierced the fkin, I made deep wounds in
feveral with a biftoury and with fciffars, and poured
into them large drops of the venom. I introduced
into the bodies of others, bits of tow moiftened
with the venom, and paffed quite through, and this
method, which I had always found mortal when
tried on other animals, was quite without effect on
this occafion, fince neither of the leeches died. I
preferved for feveral months, in glaffes filled with
water, parts of leeches quite alive. Each of thefe
pieces preferved the motions there, that it poffeffed when united with the other part of the body.
I had feveral of thefe bit by vipers, and made
notches in others, paffing into them bits of tow
fteeped in the venom : neither of them died.
They preferved all their motions, and did not feem
at all incommoded. The leech then has the property of refifting the venom of the viper, which
is quite innocent to it.

I afterwards wifhed to try what would be the
effect of the venom of the viper, on fnails and flugs.
I procured the largeft of them, and of different
kinds. I had fome of them bit in feveral parts of
the body, and by feveral vipers. I likewife made
incifions, into which I introduced the venom,
taking good care before hand to wipe off the glutinous matter that covers them, that the poifon

might

might penetrate the readier. Out of twenty-seven
flugs and fnails on which I made thefe experiments,
one flug only died, twenty hours after it had been
bit. I could not even fucceed in killing them with
the envenomed bit of tow introduced into their
bodies. The greater part of them covered themfelves with their vifcous flaver on being bit.

In the country about Pifa a fnake is found, which
the peafants call afpick, and which they reprefent
as more venomous than the viper. This creature
has fome exteriour refemblance to the viper, but
has neither the canine teeth, the fheath, nor the
veficle or refervoir of the venom ; and the experiments I have made, have convinced me that it
is an animal in no way dangerous. The fnake with
two heads that was brought to Redi, and of which he
gives a defcription at the beginning of his obfervations *on living animals that are found in living animals*,
was of this fpecies. That of Redi was fingular
however in having two heads. I wifhed in the firft
place to be certain whether the venom of the viper is mortal to this kind of fnakes, and had one
of them bit twice in the tail by a large viper.
Two days after, I had it bit by two others in the
back, from which a little blood flowed, and after
two days more held three other vipers to it, which
gave it feven or eight blows on the neck with their
teeth. It was rendered a little torpid by them, and
was flower in its motions, but on examining it two
days after I found it alive, and on putting it to the
ground

ground it crawled along and feemed in perfect
health.

The venom of the viper has no greater action on
another larger fnake, which in Tufcany is particularly diftinguifhed by that name; this is the *adder*.
I had feveral of them bit on the back, tail, neck,
and belly; to fome of them I have even held three
vipers at a time: neither of them died.   They
did not even feem furprized at it, neither were
they benumbed.   I at length employed the envenomed tow, forcing it into wounds I purpofely
made.   In fome of them I even removed the fkin
at particular parts, to convey the venom the better
to the blood.   All thefe attempts were without
effect.   It feems certain then that the venom of the
viper is in no way either mortal or dangerous to
this fpecies of fnakes.   It is not alone then on animals of the *worm* clafs that the venom of the viper is deftitute of action; there are others again,
the organization of which is more compofed, and
which have a heart and many vifcera, and are notwithftanding out of the reach of its fatal influence.

I have found another fnake called cecilia, the
orvai of the French, which likewife refifts the bite
of the viper.   I have frequently made experiments
on thefe fnakes, and have had them bit by feveral
vipers at a time, and in different parts of the body.
This creature, naturally fluggifh, did not feem incommoded by the venom, even when I introduced
it into its body by the means of incifions.

D 4   Thefe

Thefe three fnakes, the afpick, the cecilia, and adder, are not venomous, fo that one incurs no rifk, even when they bite fo as to draw blood; they have no canaliculated teeth, no fheath fuch as that which covers the teeth of the viper, nor a refervoir for venom; in a word, they are creatures perfectly innocent, as I have affured myfelf by many experiments.

I had two turtles bit by a very large viper, rendered furious, in the hind feet where the fkin is the leaft hard. I kept them alive for more than ten days, during which time they did not appear to fuffer, and walked as ufual. I had another bit feveral times in the neck, and as a clear proof that the teeth of the viper penetrated through the rough fkin, one of them was left in that had forced its way into the vertebræ. On the day following, I had it bit in the neck by another viper, and by a third in the fore feet. Laftly, on the third day it was again bit by two vipers in the neck and hind feet. It not only furvived, but did not feem to have fuffered the fmalleft inconvenience. One would have faid on the contrary, that it was become more fenfible and active.

I had five others bit in the breaft and belly by eight vipers, after removing the inferiour fhell, and laying the flefh bare. Neither of them died. They were even living four days after, which is ufually the cafe with thofe that have been deprived of the lower fhell only. In others I made deep wounds in the feet, and even removed the fkin in

fome

fome of them to introduce the poifon the better. At length I forced into the wounds, large bits of tow foaked in the venom. Neither of them died nor had the fmalleft ailment.

I do not believe however that turtles are abfolutely beyond the reach of the effects of the venom. One of them died, after it had been bit by eighteen vipers. The blood oozed from the wounds thefe animals had made, in every part of its body. A third died twelve hours after it had been bit in the neck by three vipers only ; and a third again in the fpace of twenty-four hours, notwithftanding it had been fimply bit in the feet by two large vipers. It appears then that this venom does but rarely penetrate and diffufe itfelf in the bodies of turtles, and that its action there is much flower and weaker than in the other animals with cold blood. Thefe laft in general, die from the effects of this poifon, at leaft all thofe I have had bit, not even excepting eels, which however die later, and not till the end of eighteen or twenty hours. The other kinds of fifh are likewife killed by the venom of the viper. And laftly, the fmaller lizards fcarcely furvive its bite for a few minutes.

Animals with warm blood are univerfally killed by this poifon ; I have at leaft never met with any one that furvived its action. A fmall gofs hawk died in lefs than three minutes. In a few feconds it began to open its beak, as if it felt a difficulty of refpiration, and had an inclination to vomit. A few inftants after it fell on its breaft, and could not

again

again recover its feet. It at length died with all
the fymptoms of an extreme debility. I have generally obferved that animals with warm blood, and
the action of the heart of which is very lively, die
much fooner than the others.

There are feveral kinds of animals, very diftinct
from each other, to which the venom of the viper
is not a poifon; or if it be fo, it is but very rarely,
and that with the leaft poffible energy. There are
perhaps many others we are ignorant of, that refift
its action. I have myfelf found feveral of the fpecies
of infects and worms to which this venom is not
hurtful. I fhall perhaps fpeak more fully of them
in another work, in which I fhall treat of the remedies againft the bite of the viper.

All thefe particulars ought to render the philofopher who ftudies nature very circumfpect, unlefs
he wifhes to bewilder himfelf at every ftep; they
likewife fhow us how little truft is to be repofed in
the fimple analogy that may be found betwixt different animals, either as it regards life, or the economy of their motions. Nature does not fuffer
herfelf to be devined. Experiment alone, in the
hands of an attentive and difcerning obferver, can
fnatch from her her fecrets.

C H A P-

CHAPTER VII.

The Venom of the Viper is not Acid.

IN a small publication of Mead on poisons, printed in 1739, with the false indication of Amsterdam and Naples, the venom of the viper is said to be acid, and to change the blue colour of the turnesol to red; of the truth of which he says he is convinced by his own experience. To be certain of this, I received the venom of a viper I had just killed, on a bit of glass, forcing it immediately from the point of the tooth, by a gentle compressure of the palate. I afterwards poured this venom on a bit of blue paper, which soaked it up, but instead of becoming red, turned a little yellow, and preserved this appearance even after it was dry. It appeared extraordinary to me that so learned a man as Mead should have been deceived in so easy an experiment. I therefore took a greater quantity of venom, with which I rubbed several pieces of blue paper, and that nothing might be neglected, varied the experiment a hundred different ways. At times, to have the venom the purer, I took it immediately from the tooth, before it had touched the other parts of the mouth; and at others again, either forced a bit of cotton into the mouth of a

living

living viper at the moment of its biting, or introduced it into that of a dead one filled with the venom.
I diluted a quantity of venom in water, and wet
blue paper with it.   I tried to find whether the
mixture of the venom with the other humours of
the animal, had not deceived Mead as to the colour, and varied my experiments for that purpofe
infinitely, but in vain.   I could never turn the
paper red. It fimply took the yellowifh tinge found
in the venom itfelf.   Mead likewife maintains,
that he has feen the mixture of this liquor with
violets become fomewhat red : I have tried this,
but the event has not been the fame.   When the
venom is in a greater proportion than the firop, the
mixture does indeed become a little yellow, but
never becomes red.   I increafed, I diminifhed the
quantity of the venom ; I have taken it pure, and
again have employed it mixed with the foam of the
viper : I could never perceive any thing befides a
flight yellow tinge, and all my experiments have
only ferved to confirm me in the opinion that
the venom of the viper neither changes red the
firop of violets, nor the dye of the turnefol (a).

In the fame work on poifons, Mead maintains
that the venom of the viper is a true acid, and that
it effervefces with alkaline fubftances.   In confequence of this I took feveral fluid alkalies, fuch

(a) Doctor James is likewife of opinion that the venom of the
viper is acid, becaufe, according to him, it changes the dye of
the turnefol and firop of violets red, as other acids do.

as the fpirit of hartfhorn, and oil of tartar *per de-liquium*, with which I mixed different quantities of the venom, always pure and unmixed with the other liquors of the mouth. I never could ob-ferve the fmalleft motion nor the leaft effervef-cence, at the moment of their union. It was in vain that I had recourfe to a microfcope, I could never obferve the fmalleft air-bubble difengage it-felf; the colour remained the fame, and I met with nothing that gave me the fmalleft fufpicion of the exiftence of an acid in the venom. It muft not be thought that the rapidity of the effervefcence prevented my feeing it, fince the drop of venom was fo flow in uniting itfelf to the alkalies, that it was eafy to follow it with the microfcope, and to feize the precife moment of their perfect union.

CHAPTER VIII.

The Venom of the Viper is not Alkaline.

As authours are to be found who pretend that the venom of the viper is alkaline, and not acid, and as it is principally on the activity and fudden-nefs of its effects that they have founded this their hypothefis, I thought it advifable to confult ex-periment thereupon. I took then different acid
liquors,

liquors, fuch as vinegar, fpirit of falt, fpirit of nitre, fpirit of vitriol, and laftly, feveral acid falts extracted from plants. I united with all thefe acids a larger or fmaller proportion of the venom, but could perceive no other than a yellow colour, which appeared whenever the quantity of venom exceeded that of the acid. I armed myfelf with a good microfcope, and never found either effervefcence, motion, or air bubble, to refult from this mixture. I tried it afrefh with firop of violets, but it did not turn it green, as alkaline fubftances ufually do.

It is equally without foundation then that naturalifts pretend that the venom of the viper is acid or alkaline; and it is ftill with lefs reafon that they have contrived to explain by thefe hypothefes, the pernicious effects of this poifon. Their irrational theories are completely belied by experiment, the only guide to thofe who enter into the fearch of phyfical truths. It muft however be acknowledged, that Dr. Mead has corrected many errours as to facts, in a new edition of his work on poifons, printed in Paris in 1751, which has reached me too late. He there retracts what he had advanced on the acid quality of the venom of the viper. He confeffes that the experiments made with the turnefol and firop of violets are falfe, and that the venom neither effervefces with acids nor alkalies. This avowal prevents me from endeavouring to account for the contradiction betwixt his experi-

ments

ments and mine, and from pointing out what may have occafioned his errour.

CHAPTER IX.

No Salts are difcovered in the Venom of the Viper.

THUS have I the fatisfaction of being the firft to confirm, by experiments more numerous and more diverfified than his, the truths which Mead has difcovered, and which no one that I know of has bufied himfelf about fince him. This conformity fixes in an unvariable way the certainty of my obfervations.

In the courfe of my refearches I have examined with the moft fcrupulous nicety into the exiftence of that pungent and cauftick falt, which Mead himfelf in his laft work, and all the obfervers after him, fay they have feen in the venom of the viper (a).

Mead regards it as a neutral falt. He pretends that he has feen it floating in the ftill liquid venom, and defcribes it as formed of very fharp

(a) James maintains with Mead that he has feen thefe falts, although in a fmall quantity, in the diluted venom. They both fay, that the net-work it forms in drying, is entirely compofed of fmall cryftals.

points.

points. But what was my furprize when on examining the venom with a microfcope, I could never difcover this collection of faline cryftals which
the learned Englifhman believed he invariably
faw! I even employed, but ineffectually, the
very ftrong lens' made in England. I could find
nothing throughout befides a yellowifh and vifcous humour, without any determinate fhape, without diftinct floating corpufcles or particles, and
alike in all its mafs, as an oil of any kind appears
when viewed with a microfcope. The venom I
employed was pure and taken from the tooth alone.
I varied this experiment an hundred different ways,
and even had recourfe to the folar microfcope; I
at length fatisfied myfelf that there are in reality no
falts in the venom, and that Mead muft have been
impofed upon by fome particular circumftance.

I then recollected that I had formerly feen with
a microfcope, certain tranfparent bodies which
floated on the human faliva, and which might
eafily have been taken for falts. Indeed any one
who is not very converfant in the ufe of the microfcope, and who is not well acquainted from habit
with the fhape of the different falts that are found
in liquors, particularly whilft they are drying,
would eafily perfuade himfelf that the fmall dia-
phanous particles which float on the faliva, are
abfolutely of a faline nature. They are however
too light, too large, and not fufficiently tranfparent, to be really falts. They vary both in fize and
fhape. The direction of thefe fmall bodies is rather

ther crooked than ftraight; they have hollows and folds on their furface; and laftly, they become fhrivelled and obtufe in proportion as the faliva dries. Thus are they to the eyes of a practifed obferver, nothing more than fmall pellicles or light, plaited, membranes, and as it were the relicks of almoft digefted aliments. In reality, they difappear on wafhing the mouth well, and I have obferved on touching them with a fine and fharp needle, that they lengthen or fhrivel up like fmall bits of fkin. I have met with fmall floating bodies fimilar to thofe that are found in the human faliva and in that of animals, with the affiftance of a microfcope, in the falivary humour of the viper. I have likewife feen fome of them floating in a drop of venom I had caught on a fmall filver fpatula, put into the mouth of a viper, the palate of which was ftrongly compreffed. I then conceived how Mead was led into this errour. He certainly took the venom from the mouth of the animal, and not in the way I did, immediately from the tooth; and regarded the fmall bodies which proceeded alone from the faliva, as belonging to the venom.

It is likewife true that fmall bodies or globules, fomewhat yellow and tranfparent, are often found in the venom of the viper whilft yet fluid. This never happens but when a ftrong compreffion is made on the palate or vefiele, at which time, far from being pure, the venom flows mixed with other corpufcles fupplied by the refervoir.

VOL. I. E In

In Mead's works we likewife find an obfervation, which is repeated in the Paris Edition, and which appears to eftablifh the exiftence of thefe falts in a clear and evident way. He fays, that in examining with a microfcope the venom of the viper put on a bit of glafs, the faline particles, in proportion as it dries, are feen to form themfelves into very fine and fharp chryftals, refembling a very fine fpider's web ; and that thefe tranfparent chryftals or needles continue perfect for feveral months, fo ftrong and firm are they, notwithftanding their fmallnefs.

I took then a drop of the viper's venom, perfectly pure and free from any mixture with the other liquors of the mouth. I dried it on a piece of glafs, and viewed it with a microfcope. What was my furprife on obferving, inftead of the drop, a heap of different tranfparent bodies, of an equal furface, and difpofed with great fymmetry and regularity ! their fhape was in general quadrilateral or triangular, and their points very fharp, fo that they ftrongly refembled the net-work Mead has defcribed. Their regularity and tranfparency might at firft fight very eafily caufe them to be taken for falts, but they were too large, and arranged with too much order, not to make one miftruft this appearance. What at length fixed me in the perfuafion that they were not chryftals was, that I did not fee any of them in clufters, as they are found in other falts ; they were all diftinct, and placed at equal diftances from each other. Perfons who are accuftomed to view the falts of other fluids, muft perceive the weight of thefe laft proofs.

proofs. I now fufpected that the venom had fplit
and cracked in different places in drying, and that
this had occafioned its being thus divided on the
glafs, as happens to feveral fubftances, which when
they dry, fplit in this way into thoufands of fragments, either pretty regularly fquared, or in a triangular form, and all at equal diftances. If thefe
cracks are throughout of the fame fize, it is owing
to the fame caufe, that is to fay, evaporation, acting
at the fame time and with fame the force on the
whole furface; it is from this that it reprefents a
kind of net-work with different meshes, exactly like
the web of a fpider.

Laftly, to make myfelf ftill more certain that thefe
were not falts, but rather fo many fcales and broken pieces of the dried venom, I fell upon a new
experiment which I thought a decifive one. I dried
a few drops of the venom, in a very pure ftate, in
a fmall concave glafs; I then examined them with
a microfcope, and found them as ufual, full of fmall
crevices, reprefenting a fpider's web. It was however very clear that thefe chinks, towards the bottom of the glafs, were larger in proportion as the
dried humour had a greater thicknefs. Thefe pretended falts were no other than the fragments of
venom feparated and dried on the glafs. Thofe
that were the thickeft had little or no tranfparency.
They were of a yellow colour, like the venom itfelf
in a fluid ftate. They are fimply caufed then by the
parts of the venom retreating from each other dur-

E 2   ing

ing the evaporation; and this is even vifible. to the eye, without the affiftance of a microfcope.

But to remove all doubt and fufpicion on a matter fo important and fo generally adopted, and on which Mead has in fhort founded his hypothefis of the action of the venom carried into the blood of animals, I made another. experiment, which abfolutely proves the nonexiftence of this pretended faline net-work. I put a drop of venom on a flat and fmooth glafs, and followed it very attentively with the microfcope during the whole time of its drying: nothing occurred however fimilar to what happens to falts diffolved in water. The faline particles during the progrefs of the evaporation collect together and approach towards the centre from the circumference, forming at firft very fmall chryftals, which encreafe in bulk, from the addition of faline particles of the fame nature which unite with them. Here, on the contrary, I found nothing befides a humour which, in proportion as it dries, cracks and prefents furrows, that form the quadrilateral and triangular fragments I have mentioned. Thefe crevices, which are like the fpaces betwixt the threads of a net, appear at firft at the circumference, and proceed gradually towards the centre in proportion as the deficcation advances. But the quadrilateral and triangular fragments that fill the fpaces betwixt the crevices, and reprefent mefhes, do not encreafe here as the faline particles do in a diffolution of falt during the progrefs of the evaporation. I repeated this feveral times with a fingular pleafure. I mixed the

the venom with a few drops of very pure spring water, which I left to evaporate, and observed it patiently with a microscope, hoping in this way to discover any salts it might contain, but I was not so fortunate : no better method can however be fallen upon for this purpose.

Two celebrated Professors of the University of Pisa, Messieurs Perelli and Lampredi, were witnesses to my experiments. They very gladly honoured me with their presence, and constantly assisted me, particularly when I made my researches on the salts of the venom of the viper. They both agree that whatever reason they might have previously had to suspect their existence, my experiments, joined to a little reflection, have been more than sufficient to destroy the very shadow of a suspicion.

It must likewise be remarked, that the clefts which form when a large drop of the venom is evaporated, are much larger than when the drop is small, or when it is dissolved in water, or very much spread on the glass ; these large clefts are disposed like rays that proceed to an union with each other, towards the centre of the dried venom. The space betwixt these rays is likewise dissected by other transverse rays, which become closer in proportion as they approach the centre, and form the above-mentioned figures, besides many other very irregular ones. These transverse clefts are smaller at the circumference, are at greater or less distances from each other, and are bent into segments of a circle.

When

When the venom of the viper is viewed with a microfcope, very fmall and tranfparent particles or fpots are likewife fometimes obferved in it, which are the laft to dry.

Thus have I fully fatisfied myfelf of the nonexiftence of thofe falts which phyficians and naturalifts have hitherto admitted with fo much confidence. I have feen the theories founded on this principle, to explain the action of the viper's venom, fall and vanifh before experiment, which proves that no falt, either acid, alkaline, or neuter, exifts in this humour.

CHAPTER X.

The Venom of the Viper has no determinate Tafte, and when put on the Tongue caufes no Inflammation.

FROM the teftimony of Redi, the venom of the viper was at firft thought to be infipid, and fomewhat fimilar in tafte to the oil of fweet almonds. We however find in no part of his works that he experienced this himfelf. He feems on the contrary to have trufted in this refpect to a certain Jacques, a viper-catcher, who was venturous enough to tafte this dangerous liquor. He boafted that he could fwallow a whole fpoonful of it, and Redi tells us that he has been feen to take it feveral times.

Mead

Mead, on the other hand, affures us that he has tafted it himfelf, that he has made others tafte it, and that it is acrid and pungent; he ʃays that it leaves a fenfation of burning on the tongue for feveral hours, notwithftanding it is diluted with warm water. He adds, that a pain and fwelling of the tongue foon rewarded the temerity of him who tafted it pure. Thefe contradictions reduced me to the philofophical neceffity of tafting the venom myfelf. I did fo, but not without repugnance; and as the celebrated Morgagni obferves in his excellent letter on poifons (a), I fhall advife no one to try it in the gaiety of his heart, left he fhould happen at the time to have fome excoriation on the tongue, which is a circumftance not always eafy to determine. Here however a point was to be fettled which has divided the opinions of the moft modern and moft reputed authours.

I put a drop of the venom then on a bit of glafs, and diluted it with ten or twelve drops of water; I touched the tip of my tongue very flightly with it, and immediately felt a fenfation as it were of cold and infipidity. I waited a little, in expectation of that burning fenfation, which acid and cauftick liquors occafion, and at length withdrawing my tongue, paffed it acrofs my lips, gums, and palate, that I might better come at the favour of the venom: notwithftanding this I could find no tafte in it, except that of a very infipid liquor. I then took all the venom I could exprefs from a viper, and

(a) De fedib. et caufis morb. Epift. 49.

E 4

ven-

ventured to put it in a pure state on my tongue, the
point of which I rubbed well with it, as the most
sensible part ; I likewise rubbed my lips with it,
I found a degree of consistence and viscosity in it,
but nothing acrid, pungent, or burning ; in a word,
it had no determinate taste. It is however not so
insipid as pure spring water. There is something
in it that resembles the almost insensible savour of
the fresh fat of animals, with a very slight flavour
which one can scarcely distinguish, but which would
be pretty like that of the viper's fat, if this last
was not stronger and more nauseous.

I found no greater taste nor smell in it on drying
it, and reducing it to powder, As I could meet with
no naturalist bold enough to make the same trial,
and as a support to my opinion, I gave it to my ser-
vant, a native of Tirol, named Jacques Benvenuti,
to taste. This man, as intrepid as the one Redi
speaks of in such terms of admiration, swallowed it
repeatedly at different times, at some times pure,
and at others diluted in water, varying the quantity,
and never perceived it to swell or burn either the
tongue or the mouth, He said however that when
he took it pure and in a large quantity, the sensation he felt was very different from that excited by
oil of sweet almonds, pure water, or either acid or
sharp substances : but he could not tell in what this
difference consisted. A sensation sometimes continued on the tongue for several hours, not of pain,
but as he described it, such as is felt on taking
some.

fomething aftringent. His obfervation was juft, for I myfelf have experienced this difagreeable kind of fenfation, which frequently continued for five or fix hours, 'n the parts of my mouth where the poifon had remained a long time. If it is taken in a fmall quantity and mixed with water, it leaves no fenfa- tion on the tongue ; and this fpecies of diforder in the mouth is not felt the inftant the venom is tafted, nor immediately after, but only at the end of a cer- tain time, and it is likewife neceffary that the ve- nom be kept a long time in the mouth. I have re- peated thefe trials more than an hundred times, and have never had my tongue either fwelled, inflamed, or painful. What is ftill more, the venom when even applied to the eyes, caufes neither pain nor in- flammation. I have laid fome of it feveral times on the tunica conjunctiva of different animals, fuch as dormice, cats, and dogs, and neither tumour nor inflammation has ever fupervened in this part, which is otherwife fo fenfible to the impreffion of fub- ftances, frequently thofe that are the moft innocent. I have even introduced it into the nofes of thefe animals, without their ever betraying any fign of fuffering the fmalleft inconvenience from it.

It is certain then that the venom of the viper is in no way fimilar to caufticks, and that it is not acrid and hot like that of the bee or fcorpion, Scarcely had I put an atom of the venom of the bee on my tongue, either pure or mixed with a little wa- ter, than it ftung and burned with as much vio- lence

lence as if I had applied the ftrongeft caufticks
that chemiftry affords. The venom of the wafp and
that of the hornet are not lefs acrid and pungent
than the bee's, and the pain that each of them excites lafts a long time. I took it fometimes from
the fting, and fometimes from the fmall veficle
that ferves as a refervoir to it, and found it in both
cafes alike, and invariably productive of the fame
pain. It ftill preferves its ftrength and caufticity
after having been dried, and kept for feveral days.

It is the fame with the venom of the fcorpion ;
the white and vifcous humour it throws out by its
fting when it darts it, caufes a fenfation on the
tongue fimilar to that occafioned by the venom of
the viper, but much weaker. It is on this account
that the fting of the bee is more painful than that
of our fcorpions. Probably the venom of thofe of
Africa is exceedingly cauftick, fince it kills animals
in a very fhort fpace of time.

I afterwards made a trial of the viper's venom on
other animals, which, although they are not, like
man, gifted with fpeech, are not backward in ma-
nifefting by figns, the pleafure or difguft they feel
on eating any thing. I put then a drop of the
viper's venom into the mouth of a dog ; the creature fwallowed it with avidity, licked its lips for a
long time, as if it had met with fomething agreeable to its tafte. I then fteeped a bit of crum of
bread in the venom, to fuch a degree that it became
quite yellow, and gave it to the fame dog, at a time
when it had already fed fo plentifully as to refufe
food.

food. It fmelt to it, and inftantly devoured it, ma-
nifefting the ftrongeft defire for more : in a word,
every time a drop of venom approached its lips, it
licked it up with the greateft fatisfaction.

Every body knows that dogs, like children, are
fworn enemies to whatever is bitter and acrid, and
that they are paffionately fond of whatever is fweet
and unctuous. Hence we muft conclude, that if the
dog found the venom agreeable, it was undoubtedly owing to its fweetnefs. Thus is it abfolutely
falfe and imaginary, that it is acrid and fiery ; as it
alfo is that the tongue, on taking it, fwells, in-
flames, and becomes painful.

Mead had an idea that the venom of the viper, when
applied to the wounds of a living animal, caufed a
very painful fenfation : a natural conception to thofe
who believe like him, that it abounds in falts, which
render it hot and cauftick. He endeavours to eftablifh his opinion on an experiment he made on a
dog. This animal did not feem very fenfible of the
pain occafioned by piercing the noftrils with a
crooked grooved needle ; but when the venom entered the wound, it howled and became furious. I
made the very fame experiment on a young dog, and
it appeared infenfible to the entrance of the drop of
venom into the wound. I muft acknowledge however that I have feen a cat fhake itfelf and become
more agitated, at the moment the venom was forced
into the lips of a wound that had been made in its
nofe. But this experiment is always liable to errour, fince the needle not only remains in the
wound,

wound, but the motion of the animal is still another cause of its being more agitated there, and of
its being forced still deeper, and causing a greater
laceration of the parts. This is doubtless sufficient
to renew the pain, and even to wound the nerves
that escaped the first introduction of the needle.

I have often poured the venom into incisions
made with a lancet, and could at no one time assure
myself to a certainty, that the introduction of it was
productive of pain; although it sometimes happened that I was 'pretty much convinced to the
contrary.

But were it well proved that the venom of the
viper causes pain, does it follow that we can draw
an indubitable conclusion that it abounds with salts,
or that it is acrid and caustick ? As if we had not
examples of a juice which, though insipid to the
taste, brings on violent pain when applied to a
wound. I have myself known people, who having
been bit by a viper, have notwithstanding felt but a
very slight pain, such as might very well have
been caused by the simple blow of the tooth. We
have a dexterous viper-catcher at Pisa, named
Bongi, who having one day been bit in the finger,
did not perceive it till he saw the blood flow, a
proof that he did not feel any pain. His father relates a similar circumstance ; he likewise had been
bit in the finger, and compares the pain it occasioned to the bite of a fly. However they both in
the conclusion became very ill of their wounds, a
clear demonstration of the venom having found its
way

way into the blood. I am well perſuaded then from experience, that the venom of the viper is neither acrid nor burning *(a)*, and that it does not contain thoſe ſalts which ſo many writers have imagined, either for the purpoſe of explaining its mode of action on the blood, or becauſe they have been imperfect in their obſervations.

CHAPTER XI.

Properties of the Venom of the Viper.

THE deadly humour furniſhed by the viper, which I have neither found to be acid, alkaline, nor cauſtick, ſubſides inſtantly on being thrown into water, like certain heavy oils drawn from vegetables. In this ſituation the parts of it preſerve their viſcoſity and natural union, and remain in that ſtate for ſome time, without changing either their primitive colour, or their tranſparency. This poiſon then is heavier than water, and differs in that reſpect from common oils, from the fat of animals, and from that of the viper itſelf, all of which float on water. Oils and other liquids that are heavier than water, ſhould be at leaſt ſuſpected, and are indeed often found, to be very violent poiſons. Without mentioning the oils of the common and cherry lau-

(a) The modification this expreſſion admits of, will be ſhown in the ſequel.

rel,

rel, the red oil of bitter almonds by diftillation is a poifon.

My next enquiry has been to know whether the venom of the viper is inflammable, that is to fay, whether the phlogiftick principle it contains is capable of taking fire. I have thrown it on burning coals. I have fteeped a piece of paper and a bit of wood in it, and I have collected it in fmall drops on the point of a needle : it has never taken fire, and I have not found it to be more inflammable than the other fluids of animals.

This obfervation holds good as to the venom of the bee, and thofe of the wafp, hornet, and fcorpion, which fo far refemble that of the viper. They all confume and dry in the fire, without kindling into a flame.

If a pure and frefh drop of the viper's venom is applied to the mouth, it is found to poffefs a certain vifcofity ; but when it is dried in large drops on a bit of glafs, it has the appearance of a tranfparent and yellow jelly : it then, like pitch, adheres fo ftrongly to the teeth, that it is with difficulty detached from them.

CHAPTER XII.

Peculiarities of the Venom of the Viper, and that of other Venomous Animals.

I T has been feen that, contrary to the opinion of Redi, the venom of the viper flows from the hole at the point of the tooth, and that it enters by the hole fituated at its bafis. According to this difpo-fition one would be tempted to believe, that thefe teeth have been formed for the exprefs purpofe of killing, fo much does the fmall hole at the point feem calculated to convey the venom into the blood of the animal bitten. But I do not here pretend to recur to final caufes, and am very far from think-ing that all this fingular mechanifm in the viper, has been exprefsly made for the deftruction of other living creatures. The venomous liquor with which it is provided is perhaps neceffary to its di-geftion, and I fhall fhow that it fingularly difpofes the flefh on which the viper feeds, to a fpeedy pu-trefaction ; a degree of change 'tis neceffary for it to undergo for the purpofe of being well digefted. However, by an unlucky but neceffary mechanifm, the fame tooth at once conveys the poifon into the animal the viper bites, and into the aliments it feeds on. Who knows but that the depriving it of this venomous humour would expofe it to acci-

dents

dents fimilar to thofe that happen to other animals, from a defect or depravity of fome one of their digeftive juices?

If it were true, for example, that the human faliva, as has been believed, is a poifon to certain kinds of animals, and a philofopher in this number, reflecting and reafoning on its nature, fhould obferve that this faliva is one of the principal juices that concur to our digeftion, would this reafoning animal be miftaken? Would it not, on the other hand, have divined nature? But if, on the contrary, one of thefe fpecies' of animals fhould pretend that our faliva has been fupplied us for the purpofe of poifoning them, fince it actually does deftroy them, would it not be fallen into a very abfurd errour? See however where thofe incautioufly hurry themfelves, who inceffantly recur to final caufes, in the examination and explanation of natural facts and events.

Finally, it is a general law of venomous animals that wound either with the tooth or fting, to convey the venom into the wound by holes or orifices they have in thofe parts. As to the fcorpion, writers do not agree either on the number or fituation of thefe orifices. Redi, by an inconceivable fatality, could never difcover them; and as he had feen only a fingle drop of venom on a plate of iron againft which he had made a fcorpion ftrike its fting repeatedly, he inferred from thence that there was a fingle hole only at its extremity. Valifnieri reckons as many as three. It is however very cer-

4   tain

tain that thofe of Tufcany I have examined, never had more than two lateral openings, through which the venom flowed, and that neither a fingle one nor three are to be found in them, as thefe two obfervers have maintained. When the fmall veficle which terminates the tail of the fcorpion, and at which the fting begins, is gently compreffed, the two lateral apertures, and likewife the venom at the crifis of its flowing out of them, are feen with the help of a good magnifying glafs.

But to return to the viper; its venom preferves itfelf in the cavity of the tooth for feveral years, without loofing its colour or tranfparence. If this tooth is then put into warm water, the venom diffolves very quickly, and is ftill capable of killing animals. It befides preferves its activity for feveral months after being dried and reduced to powder, as I have many times experienced in common with Redi. It is here fufficient that it is conveyed into the blood as ufual, by the medium of fome wound: it muft not however be kept too long, fince I have frequently known it to lofe its effect at the end of ten months.

I do not hefitate to believe, that the animals the death of which is occafioned by their touching the heads of vipers, even a long time after they are dead, are in reality fimply poifoned by the venom lodged in the cavity of the tooth, which being diffolved by the blood of the wound, may have flowed out at the elliptical hole at the point of the tooth. A bit of dried venom that may happen to adhere to

the outer furface of the tooth is likewife capable
of producing this effect.   I am well affured by all
the obfervations I have made, that the head of the
viper dies in lefs than twenty-four hours, and
that its mufcles dry in a few days provided they
are in a very dry place, or foon become putrid if
the place is wet.   The teeth of the viper are befides very fharp, fo that they pierce the fkin, however flightly they touch it.   I have twice fucceeded
in killing animals by fimply wounding them with
a tooth which had been plucked from a viper feveral hours before, and which was filled with coagulated venom.   If the nephew of the famous Jaques
the viper catcher, as Redi informs us, wounded
himfelf feveral times in the hand fo as to draw
blood, with viper's teeth he had juft plucked out,
without his ever experiencing any other ill effects
than what would have refulted fromr the prick of a
pin or of a thorn, he did not however at any time
make the experiment without the greateft rifk of
there being fome remains of this mortal poifon
in the tooth.   The chickens likewife that Redi
wounded in feveral parts of the body, with the teeth
plucked from a living viper, all incurred the fame
rifk.

I do not deny but that the venom contained in
the veficle may be likewife capable of killing, even
on the day following that on which the head has
been feparated from the body of the viper.   To
effect this it will be fufficient that the animal has
not bit before it was killed, and that the head is

4   ·   neither

neither too dry nor too rotten, fince in thefe cafes, the veficle would either be deftroyed, or could no longer convey the venomous humour to the tooth, by the excretory conduit already obftructed and dried up.

From what has hitherto been faid, it may be conceived how certain mountebanks, according to the relation of the authour of a work on Theriaca, dedicated to Pifo, fuffered themfelves to be bit by vipers with impunity. " There are men," fays this authour, " who under the pretence of poffeffing " an antidote, have themfelves bit by vipers ; they " previoufly give them a certain pafte which ftops " up the holes in their teeth, and thus renders " their bites ineffectual, to the great aftonifhment " of the fpectators, who are ignorant of the method " employed by thefe people to conceal their im- " pofture." This paffage clearly fhows us, that even in thefe times they had fome knowledge of the ftructure of the viper's teeth, and that they were of opinion, that the venom was carried by this hole into the wound. We likewife find in the work of Chryfogonus, entitled *De Artificiofo modo Curandi Febrium*, that this authour, who lived a long time after, was of the fame opinion. Speaking of the viper, he fays, " it has two teeth, the right and the " left, fixed in the lower jaw, and each of them " perforated ; they are longer than the others, and " are fhed every year when thefe animals quit their " fkin : thefe two teeth are enveloped in two ve-

F 2 " ficles

" ficles filled with venom, whence it flows into the.
" tooth by the hollow canal, the inftant the viper
" bites."

This authour feems only to have added miftakes
to what was known of the natural hiftory of the viper before his time. It is falfe, for example, that
it fheds its teeth every year, when it changes its
fkin ; it is falfe that the two veficles furround the
teeth ; it is falfer ftill that thefe two teeth are placed
in the lower jaw. This alone proves fully that he
never examined the mouth of the viper.

I endeavoured myfelf to have animals bit with
impunity, and for this purpofe prepared. a pafte
with pitch, turpentine, and yellow wax. I made
two vipers bite feveral times at this compofition,
and they remained for fome days without being
able to do any mifchief. I found that their teeth
towards the point, were indeed filled with this glutinous pafte, which ftopped up the orifice out of
which the venom fhould have flowed.

I do not believe however that this method is a
certain preparative againft the bite of thefe animals.
We have feen that there are circumftances in which
the venom may likewife pafs immediately from the
excretory conduit into the fheath. The fureft way
then would be to entirely remove the refervoir ; and
thus the mountebank would impofe on the fenfes
of the vulgar with greater certainty, fince he
would no longer have any thing to dread from thefe
dangerous animals.

There

There are excellent naturalifts who believe that
the fly, which they call in Tufcany, *Affillo*, (the ox-
fly) throws out a venomous and cauftick juice from
the end of the fting it has at the extremity of its
belly. Valifnieri, who has written fo well on this
infect, thinks that when it pierces the hide of the
larger animals with this very fharp fting, it infinuates into it a fpecies of poifon of a very corrofive
nature, which irritates and as it were burns the tender filaments of the nerves of the part, fo as to produce fpafms, throws their blood into an effervefcence, and drives them to madnefs *(a)*.

Reaumur, that great and exact obferver of the
minuteft animals believes, in oppofition to the opinion of Valifnieri, that this pain is rather the effect
of a fimply mechanical wound, than of a venom or
any other cauftick matter that the ox-fly may throw
out of its fting *(b)*.

The celebrated Morgagni, after having nicely
weighed thefe two opinions, does not precifely embrace either of them, but feems to combine them
fo as to form one opinion out of two. He maintains
that the pain which the fting of this fly caufes to
animals, frequently depends on two caufes at the
fame time; that of a confiderable nerve wounded
by the fting, and of an acrid and cauftick venom
which irritates the nerves *(c)*.

(a) Tom. I. Page 229. Venezia,
(b) Hiftoire des Infect. Tom. IV.
(c) De Caufis et Sedibus Morborum. Lib. II.

There

The opportunity I had of procuring thefe flies, infpired me with the wifh of examining them. The ancients were acquainted with a fly that, with its fting, threw whole herds into fury. The Greeks called this fly *Oeftros*. The Latins have likewife mentioned a fly, the fting of which produced the fame effect on large animals. This they named *Afillus*. I do not doubt but that the *Oeftros* of the Greeks, and the *Afillus* of the Latins, is the fame with the *Tabanus* of Varro and Pliny. And although the ancients have difcovered their ufual negligence in the defcription they have given of this fly, it is however impoffible not to fee that it is no other than the *Afillo* of the Tufcans, and the *Taon* (ox-fly) of the French. We muft otherwife determine within ourfelves, that a fly which was fo common amongft the Greeks and Latins, has not defcended to us, and that its fpecies has been long deftroyed and extinct. I flattered myfelf that I could find with eafe the fmall veficle that contains the venom of this fly, and the hollow fting that conveys it, as they are readily found in the bee, the wafp, and the hornet: I was however deceived. The fting, much larger than that of the bee, is notwithftanding neither hollow nor channelled, and I could never difcover any cavity in it, either in its outer or inner part. I did not fucceed better in finding the refervoir of this pretended humour; in fearching for which the ftrongeft lens' were ineffectual; it was in vain that I compreffed the extreme part of the belly and the root of the fting, I could never perceive a fluxion

of

of this liquor, as it is feen in the bee, wafp, and
hornet ; and in a word, in all the animals that convey venom into the wounds made with their ftings.

But to leave nothing undecided on this fubject,
I endeavoured feveral times myfelf, and engaged
others in the fame trial, to difcover the venom by
its tafte, by applying the fting and the parts of the
belly moft adjacent to it to the mouth. I bruifed it
betwixt my teeth, and rolled it in my mouth, but
could not find any thing acrid or burning in it, and
did not feel the fmalleft pain or inconvenience. If
it were however true, that this humour is fo very
acrid and cauftick as to burn, as it were, the nervous
filaments of the oxen, I certainly ought to have felt
it on my tongue, fince the venom the bee carries
in its fting, caufes an intolerable fmart and pain in
that part.

It is falfe then that the ox-fly, at the fame time
that it pierces the hide of oxen, fheds a poifon. The
pain it caufes is fimply mechanical, and arifes from
the particular fhape of its fting. This is compofed
of three fmall, fharp, and pointed hooks, of a horny
fubftance, which when united together form a kind
of pincers. The ox-fly does not ufually caufe any
great pain by its fting, but if it accidentally wounds
a large nerve or other fenfible part of the animal,
or if, which is more probable, it withdraws its fting
with fear and precipitation, and in a direction oppofite to that in which it entered, it then happens
that by tearing the fkin and dragging the nerves forcibly with its hooks, it muft neceffarily caufe that

F 4   very

very violent and infupportable pain, which throws the herds into fury. We know how great a difference there is betwixt the flight pain caufed by a fharp inftrument, and that excited by a weapon that tears and lacerates the nervous parts.

I have likewife had an opportunity of examining into the nature of leeches. There are naturalifts who believe them to be venomous, becaufe the wounds they make are very painful, remain a long time open, and fometimes caufe a fwelling of the adjacent parts. But it is clearly proved that thefe fmall animals, fo ufeful in medicine, are deftitute of venom, and fimply make a mechanical wound with the very fingular weapon they have at the bottom of the mouth. This inftrument is formed by three femilunar fubftances placed at the entrance of the oefophagus, towards the centre of which their edges would meet each other, did not this cavity feparate them : they are placed perpendicularly in a direction with the length of the animal. The curved edges of thefe half-moons terminate in a horny fubftance difpofed in ridges, the diftance betwixt which gradually widening, they at length form a kind of very fine teeth, like thofe of a faw.

Thefe worms employ the following method in fucking blood. They make a forcible application to the fkin with the outer edges of their mouth. They then make a vacuum by enlarging that cavity in fuch a way that the femilunar inftrument approaches the fkin, at which time they move the three faws circularly, and by fuccefsively drawing
them

them to and from each other, they make three
notches in the fkin, which unite in a fingle point.
In proportion as thefe faws recede from each other,
the oefophagus dilates and draws into its cavity the
blood that has been pumped up.

I have tried what I advance here on myfelf. I applied a large leech to my arm, after I had cut away
half its mouth, and was enabled in this way to view
the whole of the mechanifm at my leifure.

The teeth and channellings of thefe faws are eafily feen with a good microfcope. They may even
be felt by paffing the end of the finger over them;
and by drawing the edge of a lancet acrofs them,
particularly when they have been left to dry a little,
may be heard to grate. In this ftate they may be
employed in fawing the fkin, provided they are held
firm with pincers, or turned round with their edges
conftantly oppofed to the part. I have even been
able to effect this, notwithftanding the foft parts
of thefe femilunar bodies, fuch as the mufcles, were
not yet become dry. It is eafy then to comprehend
how the leech, after having contracted and ftif-
fened the mufcles that form the greater part of thefe
faws, contrives to pierce the moft obdurate hide;
and why it is that the wounds it makes are fo very
painful, and bleed for fo long a time, fince it only
obtains this blood in confequence of having torn
with its faws, and made an opening in fo fenfible a
part as the fkin, and one fo abundantly provided
with nerves and veffels.

I here

I here conclude the experiments, which, as I have obferved in the beginning of this treatife, are the moſt certain guide to conduct us to the diſcovery and knowledge of natural truths. Facts alone are however not ſufficient to diſſipate the obſcurity that envelops them. A train of obſervations, without the help of a ſkilful hand to apply them, would be at beſt but the uſeleſs proof of a painful application. In the ſame way the moſt brilliant ſyſtems the rich and fertile imagination of a philoſoper can ſupply, do not deſerve the attention of naturaliſts, unleſs they are founded on good experiments. To come at the cauſes of the laws which regulate the courſe of the celeſtial bodies, nothing leſs was needed than the long ſeries of obſervations of the Chaldean ſhepherds, and the powerful aid of the creative genius of Newton.

C H A P T E R   XIII.

What cauſes the Death of Animals that have been Poiſoned by the Viper.

THE firſt object of my obſervations on the venom of the viper, was to diſcover the origin of the contradictions which, notwithſtanding they are atteſted on all ſides by learned men of the firſt rank, are found in the various experiments that have

been

been made on that fubject. I muft confefs, however, that in verifying and analyzing all thefe particulars, my aim has likewife been to find in their combination, if poffible, a fatisfactory explanation of the fpeedy and deadly manner in which this poifon acts,

I fhall afk then, with Redi, " in what way the " venom of the viper extinguifhes life, and brings " on death ? Whether its action depends on a la- " tent caufe beyond the reach of human intelli- " gence ? Whether, on its penetrating to the heart, " it chills and freezes up the principle of heat ; or " whether, on the contrary, by multiplying this " very principle and giving it more activity, it " kindles it afrefh and confumes it, and in this way " diffipates and refolves the animal fpirits ? Whe- " ther it acts by deftroying the fenfation of this or- " gan ? Whether, by the means of a painful irri- " tation it excites, the blood does not flow back " too precipitately to the heart, fo as to bring on " fuffocation ? Whether it ftops its motion, by " congealing the blood in its two ventricles, fo that " they can no longer dilate or contract ? and laftly, " whether it coagulates, not only the blood in the " heart, but likewife in the whole venous fyftem ?" " To refolve thefe queftions with truth," con- tinues Redi, " is a tafk I am unequal to, and I " place them amongft the infinite number of things " I now am, and fhall probably always be, ignorant " of." Other authours, bolder than he, are not afraid of expofing their fentiments, whether badly or well founded. Before I propofe mine, I think it

ne-

neceffary to relate the moft reafonable opinions that have been held by naturalifts, as well ancient as modern, on this fubject.

The learned Brogiani, profeffor of anatomy at Pifa, has written a treatife full of erudition, on the venoms of animals. He there examines, as a fkilful critick, the different fyftems and various opinions that have been eftablifhed on the mode of action of thefe poifons.

It was at firft believed that the venom, on entering into the blood, caufed a univerfal coagulation of it, precifely as acids do when they are introduced at an aperture made in a vein. The animals on whom this experiment is made, die in a very fhort time, with tremblings, convulfions, and vomitings. On opening them afterwards, their blood is entirely coagulated in the veins, and as it has likewife been found coagulated in certain animals which, after having been attacked with the fame fymptoms, died of the bite of the viper, a trifling and hazardous inference has been drawn, that the venom brings on death by coagulation. But if, according to the teftimony of Redi and the Memoirs of the Academy of Sciences of Paris, this appears not to be equally true of all the fubjects that die of this poifon; if it is likewife falfe, that they all have thefe tremblings, vomitings, and convulfions; if the blood is frequently found coagulated in this way, in every kind of dead bodies; it follows, that the queftion yet remains undecided, and the difficulty as great as before : befides, may there

not

not be other circumstances capable of coagulating the blood, and exciting the tremblings, convulsions, and other accidents, without recurring to the acid of the venom of the viper? My experiments themselves have shown me that this acid does not exist, and that no stress should therefore be laid on it.

Others have believed on the contrary, that this venom kills by exciting an universal inflammation. But how can it be thought capable of exciting it so as to occasion death in so short a time? I will go farther, and assert that the fever which constantly attends inflammation is not always found in those that die of the bite of the viper. There are even no traces of inflammation in their dead bodies, and when any such are found, it is rather the effect of some particular circumstance in the temperament, than of a proper and peculiar quality residing essentially in the venom of this dangerous animal.

The disciples of Hoffman, who, at the example of their Master, explain every thing by the atony and spasm of parts, endeavour on this occasion to avail themselves of a truth to support their opinion. They pretend that this poison excites, they know not how, an universal spasm in the machine. But again, if this spasm does not exist in all the animals that die of this poison, how can it be regarded as an universal cause? It is on the contrary certain, that they invariably die, rather from an atony and universal resolution, than from the rigidity and contraction of their members.

I pass

I pafs over feveral other hypothefes, which are nothing more than fimple conjectures, and, far from being fupported by any decifive obfervation, are on the contrary belied by experience.

I however think it incumbent on me to relate the opinion of Mead. He fets out on the exiftence of cauftick falts in the venom of the viper, and on this foundation grounds the whole of his theory of its effects. In the edition of his book on poifons, printed in 1739, we find an ample detail of the different opinions of philofophers, followed by a chain of fyftematick reafonings, which, as any one may fatisfy himfelf, are very tedious and filled with fuppofitions. His object is to fhow that thefe falts decompofe the globules of the blood, and deftroy the temperament of it; and as it is difficult to comprehend how they can in this way deftroy the whole mafs in fo fhort a time, he fays that when once the venom has infinuated itfelf into a wound, a very fubtile and very elaftick fluid rifes out of it, which in an inftant extends its action to, and brings on a decompofition of, all the parts, even the moft diftant ones, of this fluid. It is thus that a fingle fpark which touches a long train of gunpowder makes a rapid progrefs along it, and caufes an univerfal explofion, by the fimultaneous difengagement of the air enclofed by each particle.

It is without doubt unneceffary to endeavour to combat this fyftem, fince thefe pretended falts do not exift in the venom of the viper, and fince nothing is falfer than the idea of thefe fmall globules

bules of blood filled with elaftick air. It is
befides certain that the venom does not alter the
fhape of thefe globules, which when they are obferved with a microfcope, are found to be exactly
the fame as before, that is to fay, obfcure and dark
coloured at their circumference, and more tranfparent, at the centre, as fmall round bodies generally are when viewed with a microfcope. I cannot conceive how Baker, otherwife very exact in
his obfervations, could fay in his *Treatife on Microfcopes*, that the bite of venomous animals, or even
an atom of their venom, corrupts the whole mafs of
blood, and alters the folidity and fhape of the red
globules that compofe it.

It is not on this occafion alone that a belief has
been held without any foundation, of the change
of fhape of the globules of blood. The fmall rings
that have been endeavoured to be fubftituted to
thefe globules, are a proof that the light, the
microfcope, and the obferver who relies upon appearances, are frequently the fource of the pretended changes that do not in reality exift. I fhall
fhow in a fmall diftinct work (a), that all fmall
globular corpufcles, viewed with a microfcope,
feem to be fhaped like rings, becaufe the rays of
light meet the eye of the obferver in a greater
number, from the centre than from the edges.

(a) The work announced here was printed fome years ago
at Lucca. It is entitled, *Offervazioni fopra i Globetti del
Sangue.*

The

The decompofition of the globules of blood, fo frequently advanced by phyficians, is one of the rareft phenomena in the animal economy.   The phyficians who are mechanicians, fuppofe that the globules of blood are fo many fmall round veficles filled with a very elaftick air enclofed in a fine membrane; they likewife believe that thefe globules (*a*) may eafily crack and alter their fhape, even from much flighter caufes than that of the action of a cauftick falt : but the fact is, that they are not veficles, as they have perfuaded themfelves, and that they very rarely alter their fhape.

Convulfions themfelves, that are fcarcely ever felt by animals with cold blood, do not prove that the venom of the viper contains cauftick falts, the invifible points of which prick the nerves and irritate the mufcular fibres.   Narcoticks and opium bring on convulfions, but muft we therefore believe that they act by like mechanical agents?   Still more, convulfions are not always the effect of an irritating *ftimulus*; they rather arife from the deftruction of equilibrium betwixt the antagonift mufcles.   Weak languifhing animals, that die from a lofs of blood, perifh in dreadful convulfions; and yet there are in this cafe neither points nor irritating falts.   The convulfions are here likewife un-

(*a*) Let it not be underftood that they are really globules; their true fhape will be feen in a work of microfcopical obfervations which I propofe to publifh foon, and in which I fhall fpeak of whatever relates to their properties.

juftly

juftly attributed to the fuperabundance of animal fpirits ; it feems more reafonable to believe on the contrary, that it is to a defect of them, or to their irregular diftribution in the mufcles, or rather to an irregularity in the circulation of the blood, that they owe their origin.

That opium caufes convulfions is owing, in my opinion, to its deftroying at different times and in an irregular way, the irritability of the mufcular fibres. It is befides certain that men and women of a delicate and weak frame are always the moft fub= ject to convulfions ; and it is not poffible to fuppofe in thefe people a fuperabundance of animal fpirits. We know that all the mufcles, even in a relaxed ftate, preferve notwithftanding a certain tenfion of their fibres, which, when they are cut, never fail to contract themfelves, and to enlarge the wound. When a mufcle becomes paralytick, it lengthens, and its antagonift then contracts the more ; which fhows that the repofe of the mufcles depends on the equilibrium of ftrength betwixt the different mufcles, and betwixt their different fibres. The powers thus balanced, deftroy and renew them-felves at every inftant, without producing any mo-tion or fenfible change. This natural tenfion of the mufcular fibres certainly depends on an equal and exact diftribution of the fluids in the whole fubftance of the mufcles. This truth is demon-ftrated in a differtation which I publifhed in the third volume of the Acts of Sienna, which was in part reprinted fome time after at Lucca, with feve-

ral confiderable additions, and which was afterwards inferted in the firft volume of my animal
phyficks. [*Phyfique Animale.*]

But if thefe mufcles do not receive the fame proportion of fluids, or if thefe fluids reach them, or
are diftributed amongft them, with an unequal
quicknefs and energy, the equilibrium of the mu-
tual effort of the mufcles is immediately deftroyed ;
the ftrongeft of them contract; and hence arife
the convulfions and violent agitations of the whole
frame.   This is the reafon why thofe who die of
an hemorrhage, as well as thofe who perifh by poifon, are feized with convulfions : for it certainly is
not probable that the lofs of blood and of ftrength
fhould bear an equal proportion in every part, in
every mufcle, and in every fibre, whilft the circulation itfelf is unequal, and the mufcular irritability is deftroyed gradually, and in a very irregular
way according to time and circumftances.

But even though it might be concluded from the
prefence of convulfions, that the matter which occafions them is acrid and cauftick, this would not
determine it to be a falt; and becaufe falts prick,
irritate, and corrode the nerves, can we fay that
they alone poffefs thefe properties ? too few experiments have been made to warrant the maintaining
this.

The convulfions fome of thofe are feized with
who have been bit by the viper, do not furnifh a
certain argument to explain the nature of the kind
of jaundice that fometimes attacks thofe who die

of

of this bite, or who ficken with the difeafe of the venom. Some authours have afcribed this jaundice to the contraction of the biliary pores at their origin in the liver, by which all fecretion of bile being interrupted, the blood becomes charged with this humour, and depofits the greater part of it in the organs of the fkin.

Others have conceived, with greater appearance of truth, that thefe convulfions, and this violent irritation of the nerves, caufe a conftriction of the biliary ducts, fo that the already feparated bile is carried into the blood, and fpreads itfelf over the whole fuperficies of the fkin. Both thefe hypothefes however are founded on a falfe principle, fince anatomy teaches us that the nerves are not irritable, and that the biliary ducts are not compofed of mufcular fibres. The firft of them is abfurd on another account, for if the bile is not primarily feparated in the liver, and afterwards returned into the blood, how can it fhow its quality and colour? I cannot conceive how very great naturalifts have brought themfelves to think that it is not neceffary for it to be feparated in the liver, to enable the blood to take a yellow tinge, and to give this colour to the fkin.

It is not fufficient that the blood contains all the ingredients of the bile, the fixed and volatile falts, the oil, and the water, to enable it to form bile. It is likewife neceffary that the organs which concur to its generation, appropriate the matter of it, and regulate the proportions; fo that the fame

fubftances

subftances which in the proper vifcus might have
formed bile, can never acquire, when mixed in the
blood with the other principles of that fluid, either
its nature or properties.   But when once it is feparated, and thrown again into the mafs of blood,
it preferves its feveral qualities in fuch a way, that
all the principles of the blood can no longer decompofe them, or break their combination.   It
may be compared to a drop of oil, which conftantly preferves its nature in the midft of another fluid,
although agitated and divided *ad infinitum* ; each
feparate particle continues to be oil as before. Thus
for example, the principles of *muft* (new wine) and
of oil certainly exift in the vine and in the olive
tree, but they only fhow themfelves in the grape
and olive.

A more appofite circumftance ftill, and one
which ruins this hypothefis, is the example of eu-
nuchs.   The partizans of it will agree, that it is
in vain for thefe unfortunate people to preferve in
their blood, during their whole lives, the principles that conftitute the femen, fince it does not manifeft itfelf by any of its effects ; they refemble women, and never have the fmell that characterizes
the male.   I will go further, and allowing that not
only the principles of the bile are contained in the
blood, but likewife the bile itfelf, it will not yet
follow that it has the property of giving a yellow
tinge to the fkin.   Animals have been known to
have a fcirrhous liver, or a very large abfcefs in
that vifcus, for a long time, without being jaun-
diced.

diced. Let us agree then, that if thofe who are attacked with the difeafe of the venom become fo, the caufe which produces this effect muft have intercepted the courfe of the bile after its feparation in the liver, without its having done any previous injury to that fecretion. I am firmly perfuaded that it does not thus pour itfelf into the mafs of humours, but becaufe its courfe is intercepted in the ductus communis choledochus, before it difcharges itfelf into the duodenum. The convulfions of the ftomach and inteftines that attack thofe who have been bit by the viper, may very readily irritate and contract the duodenum, and fo ftop up this orifice. Neither muft we be aftonifhed at feeing the fame jaundice make its appearance in thofe who have taken other poifons, fince they alfo have the fame convulfions, with a painful drawing together at the pit of the ftomach, bilious and convulfive vomitings, a contraction about the navel, and other complaints in the abdomen. It may likewife happen in certain cafes, that the bile of thofe that have been bit, may be fo attenuated and exalted, that it may even penetrate through the fubftance of the liver, and immediately make its re-entrance into the torrent of the circulation, conveying the jaundice to the whole furface of the body. It is thus that, in confequence of its being exalted in certain difeafes, it paffes through the thickeft membranes, and depofits itfelf abundantly on the colon, duodenum, mefentery, epiploon, and peritoneum, on which, as may be found by opening

dead

dead bodies, it beftows its colour. It is well known
that there are very few humours in the animal body
that corrupt fo readily as the bile; and we fhall
foon fee that it is this principle of putrefaction
particularly, which the venom of the viper conveys into animals.

But to return to the opinions of authours, as to
the immediate caufe of the death of thofe that
are attacked by the difeafe of the venom. The
celebrated De Buffon maintains, in his great work
of Natural Hiftory, that the activity of the venom
of the viper, as well as of other active poifons,
depends on thofe microfcopical animalcules, which
are difcovered in the infufions of vegetable and animal fubftances, and which he believes to be fimple
organical particles.   I can certify that nothing like
them exifts, either in the venom of the viper, or
in the other poifons, whether of the animal, vegetable, or mineral kingdoms, particularly thofe of
the laft. I have rendered myfelf very certain of
this, by experiments made with the greateft care,
and in which I employed the ftrongeft lens'.

The authour of a book, entitled, *On The Re-*
production of Individuals, [*De la Reproduction des In-*
dividus] or rather Monfieur de Buffon himfelf, afferts, that the venom of the viper, as well as all
other active and penetrating poifons in animals and
vegetables, can be nothing elfe than thefe organical particles; and he fays, that the falts Mead
obferved are precifely the fame particles carried
to their higheft degree of activity. He likewife
                                                    believes

believes that the pus of wounds is filled with thefe
moving corpufcles : but all this is without foundation. I have fhown that thefe pretended falts are
not found in the venom of the viper, any more than
the particles fuppofed to be in motion. I have
likewife examined all kinds of wounds, whether of
a good quality, gangrenous, or cancerous, and have
never been able to find the leaft veftige of thefe
particles : I could only difcover a quantity of fmall
unequal corpufcles, more or lefs round, and fwimming in a tranfparent liquor. But what appeared
ftill ftranger to me, and which is however inconteftible, thefe microfcopical animalcules are not found
even in wounds of living animals, that come of
themfelves, whilft they are always to be traced in
animal and vegetable fubftances, put to putrify in
water, and expofed to the air.

This illuftrious French naturalift has been miftaken then in all he has written on the nature and
action of the venom of the viper, and of other poifons. The acids falts of Mead which have never
exifted in nature, and the neutral falts of the fame
authour, which are not real, have been meta-
morpofed by the fertile imagination of the elegant
French writer, into what are ftill more abfurd, organical particles, endued with motion.

It is falfe, that the corpufcles which are feen
with a microfcope in continual motion, in the infufions of animal and vegetable fubftances, are fimple
organical particles, fince they are real animals. It
is falfer ftill that thefe organical particles are feen

G 4

in the venom of the viper, and in other poisons. No motion is obferved in any poifon whatever, and there is nothing that can give one the flighteft fufpicion of the exiftence of thefe particles there. It is befides impoffible that the falts of Mead can be the particles of Buffon, fince thefe falts are merely imaginary. There is no greater truth in the exiftence of thefe particles in the pus of wounds, fince there is no motion in this fubftance. It is with regret that I fee myfelf obliged to dwell on the errours of this writer, but his authority may eafily miflead thofe who can only judge after others. How many are there who judge in this way! we may include in this number all thofe who are not capable of immediately confulting nature ; who prefer hypothefis to fact, and eloquence to truth : a fevere and candid pofterity will, without doubt, be aftonifhed to find, that there have been philofophers and naturalifts in the eighteenth century, who, even in the moft important particulars, have ventured to fubftitute conjecture to experiment, notwithftanding that the latter would have been made with as much eafe, as it would have been decifive.

"Let indolent men," obferves Senac, *(a)* "feek an amufement in devifing the fecret fprings "of nature, as obfcure politicians divine and re-"gulate what paffes in the cabinets of princes : it "is a philofophical delirium that only hurts the

(a) Fraité du Coeur, page 29. Preface.

mind.

" mind. But where life is interefted, if we are per-
" mitted to form conjectures, it is for the purpofe
" of fubmitting them to the proof of experiment,
" which ought to decide."

In this uncertainty, feeing that the opinions of the
greateft philofophers are fubject to the greateft difficulties, I thought it expedient to have recourfe to
my own obfervations. Neither of their fyftems is
fatisfactory, when we confider the quicknefs with
which the venom of the viper kills animals. I could
not comprehend how creatures with cold blood,
fuch as the frog, are fo foon deftroyed by this poifon, whilft they furvive for fo long a time the lofs
of the heart, inteftines, and other vifcera, and even
that of the brain and head.

Doctor Mead, as we have already feen, afferted
with the generality of philofophers, in the firft
edition of his works, that poifons, particularly
thofe of the animal kingdom, act on the blood,
and are carried by this fluid into the innermoft parts.
But having paid attention to the quicknefs with
which the venom of the rattle-fnake caufes death,
this illuftrious naturalift has changed his opinion in
his laft work on the fame fubject, and has fubftituted the animal fpirits to the blood. He maintains
then, that the primary action of the venom of the
viper and that of other animals, is on the nervous
fluid, which, being depraved by it, produces inflammation in the organs, and deftroys the animal
machine ; fo that the difeafe caufed by thefe venoms, communicates itfelf to the whole body in

no other way than by the medium of the animal fpirits, which finally vitiate the blood, with which
they unite themfelves. The falfity of this hypothefis of Mead will be demonftrated by and by.

Nothing is lefs known than the manner in which
this poifon acts and brings on death ; if we reflect
however on the effects of opium, its mode of action
may inftruct and enlighten us a little on that of the
venom of the viper. That vegetable juice begins
by rendering an animal weak and torpid, and foon
kills it by deftroying the irritability of the mufcular fibres, as I have feveral times obferved in animals with cold blood, and as the famous Haller demonftrated a long time ago, even in thofe that have
the blood warm. The fymptoms and accidents
that follow the bite of the viper, do not differ effentially from thofe I have juft fpoken of, and may
at leaft induce one to fufpect that the venom of that
animal likewife kills by totally deftroying the irritability of the fibres.

I recollect that being fome years ago at Bolognia,
and reflecting attentively on the action of mephitick vapours, whether natural or artificial, I could
not bring myfelf to be fatisfied with all that the
different authours have written on their nature,
and on the proximate caufe of the fudden death
they caufe to animals. Some will have it to be
owing to the exceffive elafticity of the air, and
others afcribe it to the total lofs of this fame elafticity : now thefe two hypothefes are equally belied by facts, which prove on one hand, that the

changes

changes the elaſticity of the air may undergo in
mephitick vapours, is never ſufficient to kill animals ſo ſuddenly ; and on the other, that there are
mephitick vapours in which the air abſolutely loſes
part of its elaſticity. Others have conceived that
this peſtilential vapour kills by irritating the nerves
of the bronchiæ, and cauſes a criſpation and an univerſal conſtriction in the lungs, to ſuch a degree as
to cloſe the paſſage of air, and to prevent their di-
lation. Laſtly, there are thoſe again who have
conceived, that the vitriolick particles of the mephitick vapours act with a repulſive force againſt
the particles of the animal fluids, ſo that the pulmonary veſicles, deprived of their animal ſpirits,
fall into an abſolute ſtate of relaxation. It is however very certain, that even thoſe animals which live
a long time without reſpiring, and without there
being any circulation in the lungs, ſuch as frogs,
and other animals with cold blood, and the generality of inſects, in which the circulation frequently
remains for a long time intercepted, without endan-
gering life in the leaſt ; that all theſe animals and
inſects, I ſay, are very ſoon killed in mephitick
vapours. Beſides, the nerves are neither ſuſceptible
of contraction nor of irritability, and the pulmonary veſicles are not formed of muſcular fibres. It
is likewiſe certain, that there are mephitick vapours without ſulphur, ſmell, or taſte, and which
do not contain either an acid or alkaline ſalt ; and
even though they ſhould contain any ſuch, we
ſhould not comprehend the readier how they are

capable of killing fo fuddenly, animals in which life
is fo tenacious, and which the knife, the fire, the
very extraction of the heart, lungs, and all the
other vifcera, and laftly that of the brain, do not
deftroy without great difficulty. In confequence of
thefe confiderations, I formed a determined refo-
lution to make artificial mephitick vapours, and
to examine the effects of them on living animals.
I collected the vapours of fulphur in a recipient, in
which I placed a frog; it died almoft inftantly, after making a few leaps, and being violently agitated. I opened it, and found all its parts flaccid
and relaxed. The heart ftill moved, but feebly and
with great difficulty, and in a fhort time entirely loft
thefe little remains of action. I endeavoured ineffectually to irritate, not only that, but likewife the
other mufcles; neither of them would contract. I
forced a needle into the fpinal marrow, and faw
with furprize, that it no longer awakened the motion of the limbs. The colour of the blood was
changed to brown, but its globules ftill preferved
their round and fpherical fhape.

I placed two other frogs beneath a glafs recipient, into which I had introduced the vapour of a
folution of iron filings in the nitrous acid. They
died inftantly. I opened them, and found the blood
of a brownifh hue, and collected in the auricles.
The heart was no longer in motion, and was infenfible to ftimulations. The flefh was throughout
flaccid, and had likewife loft all irritability. On
pricking

pricking the crural nerves, the legs remained motionlefs.

During this period, the celebrated Doctor Veratti likewife made experiments on artificial mephitick vapours. I affifted at them myfelf, in company with other profeffors, and they proved very conformable to mine. It refults clearly from all thefe circumftances, that mephitick vapours kill animals, by deftroying the irritability of the whole mufcular fyftem. This is the immediate caufe of their action, and the reafon why thefe pernicious exhalations kill animals as it were inftantaneoufly.

About the time when the firft part of the prefent work appeared in Italian, (at Lucca in 1767) I found, as has been feen above, that artificial airs kill frogs by deftroying the irritability of the heart; and the examination of the effects that mephitick vapours produce on living animals, made me conclude, that they occafion death by deftroying the irritability of the whole mufcular fyftem. But a celebrated Phyfician (Tiffot) feems not to be of this opinion in his excellent work on the nerves. He there expreffes himfelf in this manner *(a)*, " One of the greateft modern naturalifts " thinks that factitious airs abfolutely deftroy the " iritability of the heart, and that their effects are " to be explained accordingly : but there is no

(*a*) Traité des nerfs, &c. T. i. Seconde Partie, Article des Effets des Poifons, § 218. en note

" convey-

" conveyance by which their action can be carried
" to the heart.   Fixed air, which kills when refpi-
" red, being applied in the way of injection to the
" mufcular fibres of the inteftines, revives their
" action, awakens the principle of life, and reco-
" vers fick perfons at the point of death.   Applied
" then to the mufcles themfelves, it excites their
" irritability, inftead of deftroying it."

This is not the place to fpeak in an exprefs way
of the effects of artificial airs on the living body.
I purpofe to do this in another work on refpiration,
which has been finifhed for fome time, and in
which I fhall relate the detail of the experiments I
have made on this fubject, and give my fentiments
on the caufe of the death brought on in mephitick
airs.   However in the mean time I think myfelf
under the neceffity of obferving, that the arguments
of the learned Tiffot have not hitherto been decifive; that the queftion remains in its original
ftate; and that it fhould only be decided by having
recourfe to experiment, to which an authority of
fo great a weight as this philofopher's, is but too
capable of preventing an application.

The firft difficulty Tiffot oppofes is, that we do
not know the channel by which mephitick airs deprive the heart of its irritability.

But it muft be acknowledged, that the ignorance
of one truth does not exclude the knowledge of
another; and that we may know the effects without underftanding the caufes, and ftill lefs their
manner of acting.   All human fcience is of this
                                                            nature.

nature. We know effects, of which we are entirely ignorant of the caufes; and we know caufes, of which the mode of action is abfolutely concealed.

The queftion then is reduced to this; to determine by experiment, whether mephitick vapours deftroy, or do not deftroy, the irritability of the heart; and the difficulty propofed above is of no weight, whether we know or not, the way in which this is brought about, provided the experiment be certain, and the illuftrious writer oppofe nothing which difproves it.

I do not befides fee how we can be certain that there are abfolutely no channels by which the action of thefe vapours may reach the heart.

They deftroy animals that are made to refpire them. In thefe circumftances there is an immediate communication betwixt the lungs and thefe vapours. Fluid fubftances are continually feparated from the lungs, and this vifcus may receive others, if they chance to act on it. There may be a real communication then betwixt thefe airs and the lungs, betwixt them and the fubftances that are feparated from that vifcus. But the lungs are known o receive the blood from the heart, and to convey it thither again. I do not therefore fee why the communication, or rather the action, of thefe airs on the heart fhould be impoffible.

The other difficulty Tiffot oppofes is, that fixed air, which kills when refpired, when immediately applied to the mufcular fibres of the inteftines, revives their action, and cures difeafes; whence he

deduces,

deduces, that when applied to the mufcles themfelves, it muft neceffarily excite irritability inftead
of deftroying it, and confequently cannot deprive
the heart of its irritability.

But, in the firft place, nothing is more common
in medicine, than to find fubftances which, when
applied to one part of the animal machine, act as a
remedy; inftead of which they occafion difeafes,
and even death, when applied to others. Several
medicines, particularly in the clafs of poifons, operate precifely in this way; frefh examples of which
will be given in the continuation of this work.

Electricity occafions death by depriving the
heart and flefhy fibres of their irritability, as I have
proved in my work on *Animal Phyficks (a)*; and
this fame electricity is notwithftanding one of the
ftrongeft ftimulants to the mufcular fibres that are
known. It reftores life by exciting irritability, in
the very animals in which it had an inftant before
deftroyed it. Amongft all the ftimulants that can
be employed to call the animals back to life, that
the electrical fhock has thrown into a ftate of infenfibility, a proper application of gentle fparks
appears to me the moft efficacious remedy.

In the fecond place, the application of fixed air
has a very different effect when introduced into the
inteftines, than when it is refpired. In the fiift

(a) The firft volume of this work, which I have already
had occafion to quote fo often, was printed at Florence in
1775, and is entitled *Ricerche Sopra la Fifica Animale.*

4

cafe

cafe its action is immediate; in the fecond, it feems to need the affiftance of the blood, to convey its action to the heart. Whence it follows, that its effects may be very different in thefe two circumftances.

Thefe particulars naturally led me to think, that the venom of the viper likewife kills animals by deftroying their irritability. I procured fifty of the ftrongeft and largeft frogs I could meet with. I preferred thefe animals, becaufe they are livelier than others; becaufe they die with greater difficulty; becaufe they are more irritable; and laftly, becaufe their mufcles contract even feveral days after they are dead.

I had each of them bit by a viper, fome in the thigh, others in the legs, back, head, &c. Some of them died in lefs than half an hour, others in an hour, and others again in two, three, hours, or fomewhat more. There were fome again that were not affected, whilft others that did not die, became neverthelefs fwelled. There were likewife others amongft them that fell into a languifhing ftate, their hind legs that had been bit continuing very weak, and even paralytick. In fome of them I contented myfelf with introducing cautioufly into a wound, made with a lancet at the very inftant, a drop of venom. Thefe laft lived longer than thofe I had had bit; neither of them however efcaped. I conftantly took the precaution to prevent the venom I introduced into the wound, being carried out by the blood that flowed from it. Some of

VOL. I. H thefe

thefe frogs fwelled very much, others but little, and others not at all. The wounds of almoft all of them were inflamed more or lefs. There were fome however that died very fuddenly, without the fmalleft mark of inflammation. A fhort time after thefe animals had either been bit, or wounded and *venomed* (*a*), the lofs of their mufcular force, as well as that of the motion of their extremities, was very evident. When they were fet at liberty, they no longer leaped, but dragged their legs and bodies along with great difficulty, and could fcarcely withdraw their thighs, when they were pricked with a needle, of the pain of which they feemed almoft infenfible : by degrees they became motionlefs and paralytick in every part of the body, and after continuing a very fhort time in this ftate, died.

I now opened the abdomen, and ftimulated the nerves that pafs through it in their way from the vertebræ to the thighs. I employed the ftrongeft corrofives, but could excite no motion nor tremulus in the lower extremities. I pricked the mufcles with as little effect, and thruft a long pin into the fpinal marrow, without producing any motion or trembling either of the mufcles or limbs. In none of thefe parts, all of which had died at the

---

(*a*) I thought I might be allowed this term, to exprefs in one word, that an animal, or any part of it, had received the venom, or at leaft that it had been applied to it. *Envenomed* would be the proper term, but cuftom has given it a figurative and moral fignification, which makes me afraid to apply it in its proper fenfe. In a work of fcience, the ufe of a new word fhould be permitted, to avoid tedioufnefs or ambiguity.

fame

fame time, was there the fmalleft veftige of life.
The nerves were no longer the inftrument of motion. The mufcles no longer contracted, and were
no longer fenfible to ftimuli. The heart alone in
fome few of them continued to move languidly,
and its auricles were fwelled and blackened by the
blood with which they were furcharged. This organ did not however feem to have fuffered much
from the activity of the venom. It continued its
motion, notwithftanding the entire death of the
other parts, and renewed its vibrations on being
ftrongly ftimulated with needles. This motion and
thefe ofcillations were however but of fhort dura-
tion after the death of the animal.

Perfons have been fometimes met with, who having been bit by a viper, have remained paralytick
in fome particular part of the body during life. A
fhort time ago a woman in Tufcany, who had been
bit in the little finger by a viper, became after various other complaints, paralytick throughout the
whole right fide of her body, and could never be
cured. In a word, it is certain that all thofe who
have met with this accident, complain foon after of
an univerfal weaknefs. Their mufcles refufe their
office. They become dull and heavy, have no
longer the free exercife either of body or mind,
and fall infenfibly into a kind of lethargy: fo
true it is, that this venom induces a palfy of the
mufcles, and robs them of that active property,
called by the moderns, animal irritability. In the
continuation of this work, I fhall fhow what opi-

nion ought to be held of that syftem, and the changes I have made in it.

Thus then it appears, that animals die of the bite of the viper, from their fibres lofing that irritability, which is the grand principle, both of voluntary and involuntary motions in the animal economy (a).

From thefe experiments on frogs, it feems that the venom of the polypus is very analogous to that of the viper. Scarcely has a polypus feized an earth-worm, when it perifhes, and has no longer any motion: thefe worms are known however to be very tenacious of life, and to move a long time after they are cut in pieces. Let us fay then, that the venom of the polypus (for it is one, fince it kills fuddenly, and in a very fmall dofe) attacks the animal irritability, and extinguifhes life, precifely as does that of the viper.

After having found that the venom of the viper occafions death by deftroying the irritability of the fibres, let us examine what are the changes that happen to the mufcles after they are deprived of this property. It has been conftantly obferved, that the flefh of animals lofes its motion and irritability in proportion as it has been penetrated by a putrefactive principle. We have many examples to prove that the lofs of the former invariably accompanies the firft progrefs of the latter. Mephi-

(a) The propofition I advance here is a very general one; the different modifications it is capable of will be fhown hereafter.

tick

tick airs, which deftroy irritability, likewife haften putrefaction, and the mufcles of the animals that are killed by them, become flaccid and livid. We likewife find that thofe of the animals bit by the viper become putrid in twenty-four hours. In both cafes the very principles of the elementary fibres are attacked, and the difunion of thefe occafions the lofs of their moft innate natural properties. This disjunction of parts, which the putrefaction of the mufcles invariably caufes, muft neceffarily deprive the latter of their irritability and fitnefs for motion.

I am led to think that the venom of the viper produces a fomewhat fimilar effect, and I found my opinion principally on the analogy of the other poifons. Indeed we find that the flefh of animals which has been cut with a knife dipped in the juice of napel, inftantly becomes more tender, and fitter for culinary purpofes. Travellers inform us that in both Indies, as well as in Africa, the inhabitants ufually hunt with poifoned arrows, and that in the fpace of fix minutes, or more or lefs according to the degree of the poifon's activity, thefe arrows kill the largeft animals, fuch as lions, tigers, and even elephants. They likewife obferve that the flefh of thefe animals immediately foftens and becomes tender ; an unequivocal proof that all thefe poifons equally difpofe the flefh to a fpeedy putrefaction. I have myfelf obferved the fame thing, in frogs and other animals bit by the viper. Their flefh foftens much fooner than ufual, to fuch

H 3

a de-

a degree as to crumble at the leaft touch; it feparates of itfelf from the bones, and corrupts and
fmells in a very fhort time.

If after thefe obfervations it is almoft impoffible
to deny that the venom of the viper deftroys irritability by conveying a putrefactive principle into
the flefh of animals that have been bit, and into
their fluids, we muft agree as to the inutility of
having recourfe, at the example of mechanicians,
to cauftick, ftimulating, and invifible falts, in explaining the action of this venom. Very far from
favouring this action, we know that falts are in general better calculated to fufpend and ftop it; and
I cannot conceive how naturalifts, otherwife very
enlightened men, have imagined and wrought themfelves into a belief, that the poifons drawn from
animals, and even thofe from vegetables, owe all
their activity to certain falts of this nature. Befides, we fcarcely find the fmalleft trace of falts in
the juices of fome of thefe plants, even the moft venomous of them. I have examined feveral of them
with the microfcope, and have not the fmalleft idea
of having found any falts in them, except in the
toxicodendron, in which tree, as in other plants, we
only find a few fhining globules, fwimming in a
more or lefs tranfparent fluid. What I am very
certain of is, that there does not exift in the venom
of the viper the fmalleft veftige of thofe formidable
falts, that have been fuppofed capable of killing animals the moment they are introduced into the
blood.

It

It is the facility then with which, by the help of thefe pretended falts, the action of poifons is explained, that has feduced thefe mechanick phyficians. They have conceived to themfelves *fpiculæ* throughout, calculated to difunite the animal fibres, and decompofe the humours. But what will they reply to the example of opium ? It kills by weakening, by even deftroying, the irritability of the fibres. If the virulence of this vegetable juice refides effentially in its gummy and refinous part, will they likewife fuppofe the exiftence of falts there ? Thefe hypothefes have had their birth in a chymical laboratory, and are not the refult of conftant obfervations of the phenomena of nature. We muft agree that thefe imaginary falts have been but too much abufed. There are thofe who have not hefitated to place them every where, and who have even gone fo far as to believe that they alone are capable of awakening the fenfes of tafte and fmelling, whilft nothing is lefs demonftrated than the prefence of thefe falts in fapid and odoriferous fubftances. Befides, they do not confider that falts are capable of altering their fhape without lofing their natural tafte ; how then can they likewife change their tafte whilft they preferve the fame fhape ? It is not therefore on a certain determinate fhape that their action muft be made to depend, as certain naturalifts will have it, who, when they fet about explaining the fenfations, fee nothing throughout befides edges and points; in an infinity of cafes this is not only fuppofed, but is likewife belied by experience. If there is only need

H 4 of

of awakening the fenfations in fome of our organs, why is there fo great a neceffity for thefe falts ? Cannot this be brought about without their affiftance ? Have not other particles of bodies likewife the properties of contact and mechanical ftimulus ? Are light and air falts, becaufe they ftrike the eye and the ear ? A fubftance of any kind that acts on a nerve, may drag and relax the medullary fubftance, and may either comprefs or irritate it, independent of the caufe that afterwards conveys the impreffion to the mind or brain. If all the external fenfations are reduced to a change in an organ, other bodies may then operate as well as falts. A fluid may likewife relax the tender parts of a nerve laid bare, and may equally fhrivel and dry them. There are fpirits and oils that dry and harden the flefh of animals, and irritate the nervous and mufcular fyf-tem, and notwithftanding contain no falts. In the fame way, poifons may kill without fuppofing falts throughout, in the three kingdoms. May not an action of one body upon another exift without the affiftance of wedges and points ? Can any one fay that falts are found every where where thefe figures are met with ? or that they pre-exifted in all the fub-ftances whence chemiftry at length fucceeds in ex-tracting them ? There is no more need of this, than to fuppofe that there are falts and points in camp and jail fevers, in the fcurvy, and in a word, in all putrid difeafes, where the corruption of the folids and fluids is alike general. We muft have recourfe to fomething very different from falts to explain the

deftruc-

deſtructive force of theſe hazardous diſeaſes which overturn and deſtroy the whole animal economy in ſo ſhort a ſpace of time. Their effects, and thoſe of many other diſeaſes analogous to them, as well as the ſymptoms that accompany them, are well calculated to lead one to believe that they convey a latent *virus* into the machine, which, like the venom of the viper, brings about the deſtruction and univerſal decompoſition of the ſolids and fluids. Indeed theſe diſeaſes are invariably obſerved to be attended with convulſions, great faintneſs, a proſtration of ſtrength, drowſineſs, an exceſſive ſtench exhaling from the yet living body; and laſtly, a ſpeedy putrefaction, which almoſt immediately follows death. The very ſudden failure of vital ſtrength in the whole muſcular ſyſtem, is a certain indication that the diſeaſe attacks the animal irritability, and the principle of motion in the fibres. It is only in this way that, without having recourſe to ſyſtems, and to free and arbitrary hypotheſes, we can comprehend and explain how it is that the ſeeds of death are capable of ſpreading themſelves in an inſtant over the whole animal economy.

I preſume that it will not be poſſible for the future to entertain any doubt as to the true proximate cauſe of the death brought on ſo ſpeedily by the venoms of the viper and aſpick; and amongſt the three ſpecies of the latter, principally of that called *nintipolenga zeilanica*. This aſpick kills by occaſioning a ſudden drowſineſs and univerſal weakneſs, followed by death, in the animal ſtruck by it. In

a word,

a word, it feems that all the poifons fupplied by the animal kingdom, occafion death by deftroying the irritability of the mufcular fibres, and difpofing both folids and fluids to a fudden corruption; The fame may be faid of thofe vegetable poifons, that are no fooner introduced into the blood, than they are fucceeded by death.

But of all the poifonous animals hitherto known, the polypus feems to poffefs the moft powerful and active venom. However irritable thefe creatures may be in other cafes, and difficult to kill, it fucceeds inftantly in extinguifhing the principles of motion and life in water worms. What is very fingular, its mouth or lips have no fooner touched this worm than it expires, fo great are the force and energy of the poifon it conveys into it. No wound is however found in the dead animal. The polypus is neither provided with teeth, nor any other inftrument calculated to pierce the fkin, as I have affured myfelf by obferving it with excellent microfcopes.

Let us likewife be very cautious how we believe, at the example of many naturalifts, that life confifts in general in the circulation of the blood and motion of the heart, and that it abfolutely ceafes when this circulation is interrupted. The circulation is not general in animals ; polypuffes have not even a heart or other analogous vifcus, to bring about its operations. It is proved too, that feveral animals with cold blood live a long time without heart, and without vifcera, as is feen in frogs, turtles, and feveral

veral kinds of fish and worms, in which, although
the circulation is undoubtedly then stopped, they
continue notwithstanding to live and move, and are
wrought on by their paffions as ufual, appearing
to be still fubject, to and fenfible of, the wants of
life.

I have found many animals, infects, and worms,
in which there is certainly no kind of circulation in
the veffels ; there are others, in which it is only
imperfectly carried on in fome particular parts of
the body, and not at all in the extremities. I purpofe to give all thefe particulars to the publick, in
a work I have been fome years bufy in getting ready, entitled, *Sur les Animaux Microfcopiques (on
Microfcopical Animals).*

This errour has fpread itfelf amongft phifolo-
phers, by the help of a falfe analogy they have fuppofed betwixt animals with warm blood, and thofe
with cold ; a very dangerous mode of reafoning in
phyficks, and belied at every ftep by obfervation
and experiments. A function has been obferved to
be executed in a certain way in animals with warm
blood, and it has been immediately concluded to
be the fame in all others. Thus are general laws
made, and propofitions on fo extenfive a fcale advanced, merely becaufe nature has not been fufficiently confulted. We have needed a *Tremblei* and a
Bonnet to rid us of thefe general axioms, and of the
idea of a neceffary and common law in the generation of all animals.

I can-

I cannot forbear mentioning in this place the fin-
gularity of motion of a fmall microfcopical animal,
which Lewenhoeck has named *rotifer* (wheel-polypus). All the obfervers, even the moft modern
ones, that have fucceeded him, have believed that
this animal has real wheels (*a*); but to be certain
of the contrary, it is only neceffary to place it betwixt two pieces of glafs, and then obferve it with
an excellent microfcope. 'Tis a fmall gelatinous
worm, commonly found in the earth or fand collected by rain in the tops of houfes. I have likewife
found it in other earths, as well as in waters that
have been fometime ftagnant, and more frequently
again in thofe that have a very gentle current, and
are filled with *conferva* and other aquatick plants.
This worm is divided towards its head into two
pretty large trunks, which appear like two wheels
or ftars, from the number of fmall, extremely fharp,
and fhort, branches that are attached to their circumference. They really appeared to Lewenhoeck
to be wheels of a rare mechanifm, and every one
would judge the fame, on feeing the creature put

(*a*) Great care fhould be taken not to confound what we
imagine, with what is pointed out to us by obfervation. Indeed
there have been authours who, either guided by analogy, or
puzzled to explain fo fingular a motion, have ventured to affure us that thefe wheels are not real; they have luckily faid
the truth. It muft however be agreed that we ought to obferve,
and not to divine, the phenomena of nature. Whoever gives
himfelf up to refearches of this kind, without the faithful guide
of obfervation, runs the greateft rifk of falling into errour.

them

them in motion. But a more exact obfervation at
length convinced me that they are not wheels, but
compofed of a quantity of fmall moveable arms,
formed like pointed cones, and planted all round the
two trunks. It lets fall thefe moveable arms or
rays fucceffively, and afterwards raifes one after the
other with fo much celerity, that the eye fancies
they are turning round like the fpokes of a coach-
wheel, or rather, like the branches of a wheeled
fire-work. It never moves thefe two wheels, except
when it fwims or wifhes to eat, and thefe two ftates
are invariably the fhorteft of its life. In fwimming,
it ftrikes the water with thefe arms or branches with
great celerity, refts itfelf at different periods, and
thus tranfports itfelf from one place to another.
When it eats, it, on the contrary, fixes its tail in
fome fubftance, and afterwards turns its two wheels,
giving fuch a motion to the water, that it directs
the courfe of it towards its head, fo that it prefents
to its mouth all the fmall corpufcles with which it
is filled. The velocity of the motion of its arms or
wheels is incredible; but what is ftill more aftonifhing, is the motion of its heart. This vifcus is feen
very diftinctly with a microfcope, and can never be
confounded with any other part of the animal whatever. It is abfolutely immoveable when the worm
does not play its wheels; but no fooner are they
in motion, than the heart moves too, and its action
becomes ftronger in proportion to the quicknefs
with which the wheels are agitated, fo that their
motions are always in an exact proportion. I do

not

not take upon me to deny; but that it fometimes, happens (although very rarely, and that at very long intervals) that the heart is in motion even whilft the wheels are at reft; and as the motion of the wheels is always at the difpofition of the animal, fo likewife is that of the heart. The heart then is a voluntary mufcle, depending on the will of the animal; a circumftance which is at the prefent time *unique*, having never been obferved in any other cafe. The wheel-polypus paffes the greateft part of its life then without any motion of its heart, and confequently, without a circulation of blood, or of a fluid which receives motion from this mufcle. This does not, however, prevent it from moving during the other intervals, when it creeps and trains itfelf, as worms do, amongft the bodies that furround it.

An objection may be ftarted here, that this organ of the wheel-polypus is not the heart of the animal, but rather its ftomach, fince it is obferved to move when the creature eats; and that it is altogether extraordinary to fuppofe, that the heart is a mufcle fubmitted to the will, whilft it is not fo in any other animal. It muft be confeffed, that this is not impoffible, but it is not, on that account, very probable; and even though it fhould be true, it would be likewife true, that an organ fuch as the ftomach exifts with a voluntary motion, which, any more than in the other cafe, is not obferved in any other animal.—Thus the difficulty I encounter is of no weight, fince it muft always be agreed,

that

that a mufcular organ exifts in this animal, which, in oppofition to thofe of all other animals, is fubordinate to the will. This is precifely what I wifhed to prove by my obfervations, and my difcovery, therefore, ftill remains fuch. It is likewife to be obferved, that the rotifer puts this fingular organ in motion, even when it does not eat; that is to fay, at a time when it can make no ufe of it, provided it be its ftomach. This happens every time it fwims, or wifhes to pafs rapidly from one place to another: it has then occafion to move its two wheels, and this organ moves in confequence. Hence we fee, that the animal does not move it to eat, but that the motion of it neceffarily takes place when it plays its two wheels, whatever may be its motive for fo doing.

But fince it is certain that the voluntary motions of the mufcles of animals with cold blood, do not depend more on the circulation of humours, than does the irritability of the fibres, which feems to be the fource and principle of life and motion in the animal, it follows that, in animals, life confifts in the action of their mufcles and parts; for the moment this motion ceafes, the animal ceafes likewife to live; and its body then, as to life, differs no longer from the ftate of any foffil or vegetable fubftance whatever; and all this affemblage of veffels, fo many different organs, and this aftonifhing ftructure of its parts, are no longer of any ufe to the animal, and fhould be regarded as if no part of them any more exifted: motion being once at

an

an end in the machine, fenfation and life are fo too. The animal will return to life as foon as its parts regain their former motion, inftead of which, it dies for ever, when, as happens to man, its parts not only lofe their actual motion, but likewife the faculty of recovering it in the fequel. Thus the microfcopical eels that are found dry and withered in fmutty wheat, recover motion and life as foon as they are wetted with a little water, and again be-come lifelefs and dry, whenever they are no longer moiftened. I have repeatedly affured myfelf of this with an extreme pleafure. Thus, then, do they preferve the power of reviving and refufci-tating effectually, by the fimple prefence of the water with which they are moiftened.

The celebrated Bonguer, in his work on the fhape of the earth, relates to us, from the tefti-mony of Father Gumillo a Jefuit, and alfo of the In-dians of Peru, that a large venomous fnake is found in thofe countries, which being dead, and dried in the open air, or in the fmoke of a chimney, has the property of coming again to life, on its being ex-pofed for fome days to the fun, in a ftagnant and corrupted water.—It were to be wifhed, that fuch a naturalift and philofopher as Monfieur Bonguer, could have verified by his own proper obfervation, a fact fo important in itfelf, and rendered ftill more fo by the fize of the animal.

I have dried the worm called *feta equina*, or, ac-cording to Linneus, *gordius*, feveral times in the open air, without leaving it there too long : it

had

had loſt almoſt all its bulk and weight, and was become like a bruiſed and dry ſtraw : its ſkin had ſhrunk ſo as to leave no ſenſible cavity, and it had no longer any ſign of life or motion. I returned it into water, where, in leſs than half an hour, it recovered its bulk and weight, and ſoon afterwards diſcovered unequivocal and permanent ſigns of life.

The wheel-polypus I have ſpoken of above, likewiſe loſes, when dried, every kind of motion and life, and recovers both one and the other when again put into water. Laſtly, I left it, by way of experiment, in a very dry ſoil, and expoſed, during the ſummer, to the whole heat of the ſun, for the ſpace of two years and an half. I afterwards returned it again into water, where, at the end of two hours, it recovered life and motion. I put one of them on a bit of glaſs, which I expoſed, during a whole ſummer, to the noon-day ſun : it there became ſo dry, that it was like a piece of hardened glue. A few drops of water did not, however, fail to reſtore its motion and life.—I have ſince found a number of other ſmall animals, either on the tops of houſes, in earths, or in water, which, in the ſame way, alternately loſe and recover the uſe of their organs, on being dried, and afterwards returned again into water. I purpoſe ſpeaking of theſe little prodigies in a work apart, to be entitled, On the Life and apparent Death of Animals (a).

But it is not the ſame as to the loſs of irritability in the muſcles of the animals poiſoned by the viper.

(a) De la Vie et de la Mort apparente des Animaux.

Theſe

These remain flaccid, and their motion is lost for ever. It appears very certain, that the venom of this animal differs but little from opium as to its effects; and that its mode of action on the fibres comes very near to that of this vegetable juice. Both of them excite violent convulsions and vomitings. Each of them conveys an universal debility into the organs. They render the muscles paralytick, make the animal heavy, and finally bring on a speedy death, by destroying the irritability of the fibres. The heart alone in both cases still preserves its irritable quality for some time after the death of the other parts. It avails nothing here to animals with cold blood, that they are endued with an obstinate life, and are capable of preserving that, as well as motion, after they are cut in pieces. If either of these poisons attacks the principle of their motions, and destroys the irritability of their muscles, they die speedily, all motion is annihilated in them, and their parts will no longer give any sign of life. Their body, it is true, will preserve its organization; but an organized body that has lost its motion, is truly a body without life.

It is evident, then, that neither of the numerous hypotheses naturalists have invented, a great part of which I have taken care to relate, explain in a reasonable way the death of the animals poisoned by the viper; but that its venom kills in no other way, than by destroying the principle of motion, the only source of animal life, in the different parts. I am the more attached to the opinion that the venom of the viper acts in no other manner than by destroying

ftroying the irritability of the mufcular fibres, from
having already fhown, in a memorial printed in the
Acts of the Academy of Sienna, that the nervous
fluid is by no means the true, the efficient caufe, of
mufcular motion; and even though I fhould be of
another opinion, and fhould regard the animal fpirits as the caufe of irritability, and the true principle of all the motions in the animal economy, my
difcovery of the proximate caufe of the death of
animals that have been bit by the viper, would lofe
no part of its importance; for whether it operates
immediately on the nervous fluid, or on the mufcular fibres, it is not lefs true that this venom kills by
depriving the animal of all motion, and the mufcles
of the power of contracting.

Unlefs I am deceived, I have now, I think, hap-
pily terminated the controverfies that have fo long
kept people at variance, on the mode of action of
the venom of the viper. I believe I have explained
how it is able, in fo fhort a time, to kill even the
animals that are the moft obftinate in dying. When
once this poifon is introduced into the blood, it deftroys the irritability of the mufcular fibres, the
fource and principle of all the motions, not only
during the life of the animal, but alfo after its
death. I call every animal dead in which there are
no longer any exteriour figns by which we can fay
that it lives; and, in truth, it is only according to
our fenfes, and the information they give us, that
we can judge of the true death of animals; that is
to fay, of the precife inftant when they ceafe to

I 2 exift,

exift, and are no longer alive. Indeed, how can we
conceive a living being, without the idea of fome
motion in its organs ? We fhould otherwife introduce into phyficks a fenfelefs pyrrhonifm, and caft
trouble and uncertainty on the moft certain and moft
received notions and ideas. A principle of corruption penetrates into, and fpreads itfelf in, the folids
and fluids, relaxes and decompofes the mufcular
fibres, and caufes them to lofe the power of contracting. It is to this general law of putrefaction,
then, and to this univerfal principle of diffolution
and death, that the entire action of the venom of
the viper on organical bodies is reduced; and to
this we muft confine ourfelves, fince in effect what
is called the fcience of nature has its bounds, and
fince it is not permitted us to go beyond them.
Whatever this fcience may be, if it is true that a
putrefaction exifts in nature, and that it effects the
deftruction of all organized bodies, it is likewife
certain that we are entirely ignorant of its mechanifm. Indeed when thefe bodies are fubmitted to
its action, who is capable of informing us what is
its mode of working on them, with what power it
operates, and laftly, what changes and what revolu-
tions it caufes them to go through ? The prodigi-
ous number of fmall movements that are exercifed
on parts of an infinite minutenefs, are too obfcure
for us, and efcape our fenfes. It is, however,
enough, to fee that a general principle of putrefaction and deftruction, which decompofes organical
bodies, and gives them up to death, reigns throughout

out nature.. To feek an acquaintance with nature, we have nothing more to do, than to affemble the effects or particular accidents of bodies, and to compare them with other more general effects, that are called *principles, or laws, of nature.* This alone is what the great Newton did, when he fubmitted all the celeftial motions to the general law of gravitation. Of what confequence is it befides to an aftronomical obferver, to know the caufe of the reciprocal tendency of the bodies moving in the fky? Such a knowledge would be rather an object of curiofity to man, than a real advantage to aftronomy,

These were my fentiments when I publifhed this Firft Part in Italian, thirteen years ago. I have at prefent made but very few changes and additions, becaufe the fucceeding parts are, rigoroufly fpeaking, merely a fupplement, ferving to correct what is previoufly advanced; and becaufe order would then have obliged me to give confequences, which could not eafily have been comprehended, till after a general idea was formed of the fubject.

Deftroyed irritability in the living machine was what I moft conftantly obferved at that time. It is for this reafon that I have reduced the action of the venom of the viper to this general principle, and have entirely excluded the nervous fyftem. 'I muft confefs, however, that the number of my experiments was then very limited, and that I had not varied them fo much as I have fince done. I was likewife ignorant of the faculty of the poifon *Ticu naas*; and alfo of the furprizing effects of the oil of

I 3   the

the Cherry Laurel, for the greater part unknown to other obfervers.

I have likewife, in this Firft Part, paffed in a very curfory way over feveral other fubjects, and have even given into a few fuppofitions, which I fhall dif-cufs in my *Microfcopical Obfervations (a).* The prin-cipal objects of thefe obfervations will be, the figure and properties of the globules of blood; the ani-mals that are capable of dying and returning to life, which will give me occafion to enter into a com-plete hiftory of the celebrated eels found in fmutty wheat; and, laftly, the caufe of the death of animals furrounded by artificial and unrenewed airs.

(a) Obfervations Microfcopiques.

P A R T   II.

C H A P T E R   I.

On the Source of many Errours.

THE *ignorance* of a truth in phyſicks may hide
from us the cauſe of a natural phenomenon ;
but *errour*, when it ſupplies the place of truth, ſtops
the progreſs of ſcience, and ſubſtitutes dreams and
chimeras to facts and nature. It is at all times a
misfortune to be ignorant of a truth ; but when we
are ſenſible of our ignorance, we may ſtill hope to
come at a knowledge of it. The book that of all
others would be the moſt uſeful to mankind, is ſtill
wanted. It is one which would at once determine
what we really know, and what we do not know,
although we have perſuaded ourſelves that we do.
Our reaſonings would no longer have hypotheſis
and errour for their baſis, and inſtead of fabricating
ſyſtems, we ſhould endeavour to prepare materials.
Nature would be more conſulted ; we ſhould reaſon
leſs, and know more.

There are errours and truths which concern man
more eſſentially than others do. Theſe are, parti-

I 4

cularly

cularly, thofe which relate to the prefervation of
his fpecies.

Man is naturally fubject to certain difeafes;
whilft others again are accidental to him. Medicine employs itfelf in each clafs, and in feeking
the remedies, renders itfelf ufeful to fociety.

Thofe who have diftinguifhed themfelves in this
refearch, cannot be fufficiently piaifed. Pofterity
will do juftice to their labours, and immortality is
fecured to them. But, on the other hand, who does
not fee the mifchief that a remedy propofed in very
violent difeafes may be productive of, if, inftead of
being falutary, it is entirely ufelefs, or even hurtful ? To pafs flightly over fo important a matter,
would be expofing mankind to the worft of mif-
chiefs. The more certain we think ourfelves of
the remedy, the more we defpife the danger, and do
not endeavour, fo much as we ought to do, to
guard againft it. The difeafe attacks us, we neglect the affiftance of art, and frequently become
the victims of our own credulity, and of the ignorance of others.

The perfuafion we are in that a difcovery is made,
blunts the ftimulus that would have pufhed our refearches a greater length, and we remain for ages
in a pernicious errour, which the hope of reward,
and the afpiring to fame, would have refcued us
from. The hiftory of human difcoveries is filled
with examples of this. We owe every thing to
thefe two great fprings of human actions, intereft
and ambition. But when a perfuafion of knowledge

is

is entertained, all inveftigation is laid afide; and
far from any further difcoveries being made, the
very hopes are loft of knowing any thing more.
Such was the deftiny of Europe when in a ftate of
barbarifm and ignorance; and fuch are ftill the
ideas of the favage.

It is now more than ten years ago that I publifhed a work *On the Venom of the Viper*, in Italian.
It is this work which forms the firft part of the
prefent treatife. I then in a manner engaged myfelf
to the publick to give a fecond part, in which I propofed not only to fpeak of the remedies againft this
venom, but likewife to treat of feveral other interefting points altogether new. I had neither time
nor convenience to finifh all the refearches I had
then in view. I was defirous of certain and evident
confequences, and it was neceffary to multiply experiments infinitely, and to vary them a thoufand
ways. But the little fuccefs I met with in feeking
a fure remedy againft the bite of the viper, particularly occafioned me to defer the publifhing the Second Part for fo long a time. Not but that I tried
a great number already known, and feveral others
which either imagination or chance threw in my
way; but they all appeared to me more or lefs ineffi-
cacious, and I found none of them certain. It is natural to fuppofe, that, amongft other remedies, I did
not forget that moft famous one of all, *Eau de Luce*;
(which indeed is nothing more than the *fluid volatile alkali*, joined with a little oil of amber, that does
not at all alter its qualities) I tried it, but the fuccefs

cefs by no means anfwered my expectations, fo that
I at length abandoned it, as I had done all the reft.

A new work has lately awakened the attention of
the publick on the advantages of *the volatile alkali*
againft the venom of the viper *(a)*. It is announced
in this work, in a tone of affurance and perfuafion,
to be *the true fpecifick* againft this dangerous poifon,
as well as againft almoft all difeafes, however dreadful. In reading this work, I apprehended that I
had been miftaken from beginning to end. Indeed
when I reflected on the experiments I had made in
Italy, I knew no longer what to believe, and was at
times even led to think, that the vipers of France
are either lefs venomous and deftructive than thofe
of Italy, or that they are of a different kind; fo
true it is, that felf-love does not allow us to acknowledge our errours till the laft moment !

But what furprifed me ftill more, was to fee the
errours of Redi, on the ufe of the bag that covers
the canine teeth of the viper, which were refuted
more than thirty years ago by Mead, again make
their appearance in modern publications ; to fee
likewife the errours of Mead on the acidity of the
venom of the viper, which he himfelf had abjured ;
and, laftly, to find thofe of the fame authour on the

(a) This work is entitled; *Expériences propres a faire connoitre
que l'Alkali volatil fluor eft le remede le plus efficace contre les afphyxies. Paris.* The Authour is Mr. Sage of the Academy of
Sciences.

faline

faline nature of this venom, which have been refuted
in Italy for more than ten years *(a)*.

If on one hand I could not perfuade myfelf that I
had been deceived on fo many points and queftions,
which I had, however, examined without prejudice, and with a wifh to difcriminate nicely ; on the
other hand, it was impoffible for me to conceive,
how certain authours could advance fo many matters as facts, with fo much affurance, and without
having previoufly convinced themfelves by certain
and repeated experiments. I could not comprehend any more, why the authours of thefe new publications had neglected to demonftrate clearly the
fource of the errours into which the writers pofterior to Mead had fallen, who had flattered themfelves that they had fhown with the ftrongeft evidence, both by obfervation and certain experiments,
how Redi and Mead had each of them been led into errour.

As the publick are perfuaded, that in phyficks,
fubjects are fubmitted to experiment, and not to
authority, thefe gentlemen ought to have expofed

(a) There is however, nothing aftonifhing in this, when we
confider the mode that is generally adopted by our modern writers. More than two hundred authours may be named who
have copied from each other on this fubject, and have given us
grofs errours for demonftrated facts. One might reafonably ex-
claim to them, " Modern parrots, the copyifts of other parrots,
ceafe to deceive us, and for once confult nature. Had you employed the time you have fpent in copying each other, in making
experiments, how many errours, and how much time, would you
have fpared to pofterity !"

experiments

experiments to experiments, and obfervations to
obfervations, and to have developed the errours into
which we are fallen. But they have not made any
fuch attempt. They have fubftituted their authority to experiment, and their name to obfervation.
This method is altogether pernicious, and neceffarily tends to perpetuate errours amongft men, and
to render difputes eternal. When we know that two
obfervers do not agree on a fact, or on an experiment, to which of the two muft we truft, provided
they are both of an equal merit? We remain in an
abfolute uncertainty, and can only have acquired,
on their perufal, a reafonable pyrrhonifm.

But is there no touch-ftone to enable us to judge
where the miftake lies betwixt them, and of two
contradicting experiments, to diftinguifh the true
one from the falfe?

The difficulty of judging betwixt two authors,
even in matters of fimple fact, has been the occafion of many errours and hypothefes having lafted a
long time, even after their falfehood has been demonftrated; and many truths have been rejected, merely becaufe experimenters have not been able to repeat the experiments that proved them, in the fame
way in which they were firft made.

For my part, I think it the duty of the lateft obferver, not only to repeat faithfully the anteriour experiments that contradict his, but likewife to give
his own in fuch a way, that they cannot leave the
fhadow of a doubt in the mind of the reader.
Without this provifo, he will lofe the aim he pro-
posed

pofed to himfelf in writing—that of being believed; which he will not deferve, although he may, by accident, have faid the truth.

There are three principal methods of avoiding this inconvenience, which perpetuates errours, and ftill keeps us in a very dangerous fcepticifm.

The firft is, to multiply the experiments exceedingly. It is almoft impoffible, in repeating them fo many times, that fortuitous cafes do not occur to vary them, and that the final refult of fo many of them is not certain and conftant.

The fecond is, to vary them in a thoufand ways, changing the circumftances as the nature and fpecies of them may require, and giving them all the precifion and fimplicity they are capable of. This method fuppofes much greater talents and genius in the obferver than the firft, and there are few of thefe, even amongft the moft fkilful, who can boaft of having invariably put it in practice.

The third method is, not only to fucceed in making experiments, decifive by their number, variety, and fimplicity; but likewife to attain to a difcovery of the fource of the errours that others have fallen into.

It is a fault, then, in thofe who write the laft, not to enter into a very minute detail of their experiments, and to endeavour to demonftrate their fuperiority and exactnefs, in comparifon with thofe of their predeceffors. It, however, is particularly incumbent on them to trace the origin of errours, and to fhow how the former obfervers have been deceived,

ceived. Without this, all their labour is a pure
lofs, and they are by no means worthy of confidence.

From all thefe confiderations, I have deemed it
neceffary to return to the fubject of the prefent
work, and to treat it in as particular a way as my
circumftances will allow me. The importance of
the fubject requires this of me, fince it regards a
very dangerous and mortal difeafe, which impreffes
with fear thofe who have the misfortune to be attacked by it, and creates uneafinefs in families.

Perfuaded that a perfect knowledge of the venom of the viper cannot be acquired unlefs by a
fearch into all its properties, which are, in a greater
or lefs degree, unknown, I wifhed that neither of
them fhould efcape me, without fubmitting it to a
rigorous, and, at the fame time, impartial, inveftigation; and that nothing which related to the fubject fhould be wanted, was defirous of examining
afrefh the fuppofed acidity of this venom, and the
falts of which fome people will have it to be compofed.

Any errour whatever that relates to this fubject,
may, in time, become dangerous in its tendency. Authours, perfuaded by a miftake of Mead,
that they were acquainted with the true nature of
the venom, have been ready to fabricate fyftems to
explain the way in which it acts, and how and by
what mechanifm it is that it produces fo fpeedy a
death. They have afterwards invented remedies
that relate to the fuppofed nature of this poifon,

and

and what is still more strange, have found them efficacious. They have shouted victory, both on occasion of the theory and the remedy, and have shown how the one served as a guide to the attainment of the other. In a word, they have pretended that all is done, and that nothing more remains to be known of the viper's venom; maintaining, that they are acquainted with its nature, its mode of action on the animal machine, and lastly, with the remedies capable of destroying its effects.—But let us leave these authours with their sectaries, to applaud themselves on knowing so many things, and on having divined nature. I believe, on the contrary, that we as yet know nothing about it, and that this matter is altogether new. My experiments will show this, in the course of the present work.

A great part of these experiments required the assistance of several persons, and I have reason to congratulate myself on this necessity, since, amongst others, it procured me the presence of two men of rare talents; Dr. Troja, Member of the Royal Academy of Naples, author of several excellent tracts on animal physicks, who happened to be at Paris at the time I made my experiments (a); and M. Jean Fabroni, of Florence, a fellow traveller, and attached to the cabinet of natural history of the Grand Duke of Tuscany, a well instructed, and very promising

_____

(a) M. Troja visited me almost every day, to observe my method of making experiments on various subjects in physicks.

young

young man *(a)*.—I name thefe gentlemen here
with the greater pleafure, fince, in thus publickly
teftifying to them my gratitude and efteem, I give
my experiments a greater degree of authenticity.

The firft queftion I now undertake to examine,
and which has been the principal occafion of my enquiries, is, whether the *fluid volatile alkali* is a certain remedy againft the bite of the viper; that is to
fay, whether it refcues from death an animal that
would otherwife have perifhed by it. This firft refearch is clearly very interefting, and deferves to
be examined with all poffible attention. I have
multiplied my experiments on this firft point, in a
way that more than one of my readers will deem unneceffary. But I know of what weight the prejudice for a favourite hypothefis and the authority of
a celebrated writer, are. Errour and truth feem to
meet with the fame difficulty and refiftance from
mankind ; one in unrooting, the other in eftablifhing, itfelf. The Newtonian fyftem was combatted
for a whole age before it was received, and it required as long a time to abandon that of Defcartes.
What is very certain is, that fo many errours have
not been fpread abroad, as to the nature of the venom of the viper and its remedies, but becaufe too
few obfervations have been made, and experiments
too little diverfified.

(a) M. Fabroni was likewife prefent at the experiments I made
in London and on my return into Tufcany, and willingly
charged himfelf with the defigns of the plates in this work.

Mead

Mead himfelf was not exempt from this fault, as I fhall fhow in examining the remedies he has pro- pofed againft the bite of the viper. The ufe of the volatile alkali itfelf was only introduced in confe- quence of a falfe theory on the nature of the venom, and was only fupported with fo much prejudice and obftinacy, for want of the making of a fufficient number of experiments. 'Tis on the fame account that the difputes on animal phyficks, which would have terminated at their birth if experiments had been much more multiplied than they were, ftill exift. But the art of experimenting is flow and painful, inftead of which it cofts but little trouble to follow the authority of another. It is eafier to reafon than to make experiments; and this art, inva- riably long and difficult, is not within the reach of every one.

Other readers will find, that the number of my experiments, however great it may be in itfelf, is not fufficient to decide all the queftions I examine in this work, nor to terminate all the refearches I make into the venom of the viper. I have nothing to oppofe to thefe laft, and, likewife, I do not take upon me to fay, that all the confequences I have deduced from my experiments are certain. Per- haps a number of experiments twice as great, would fcarcely be fufficient for this. Thofe who are acquainted with the difficulties that are met with in experimenting on living animals, and who know how much the circumftances betwixt one animal and another vary, (which rigoroufly fpeaking, are never the fame) will agree with me on this head.

Let all that has been written on the irritability and fenfibility of the animal fibres be examined, and the fame inconveniences, the fame difficulties, will be difcovered. It is true, that a very great number of experiments have been made in a few years, and that an infinite number of animals have beeen facrificed to philofophy, or publick utility; but much remains yet to be known, precifely becaufe the number of experiments is not yet fo confiderable as it ought to be.

I muft likewife confefs, that I have wanted both time and patience to do more. Nothing but the idea of publick utility can fupport the horrour of feeing fo many animals, fenfible of pain like ourfelves, fuffer under our hands; and to view them expofed to a thoufand kinds of torments. I leave the purfuit of this career to thofe who are more courageous than myfelf. The road is open to obfervers, and I fhall rejoice to fee them embrace with ardour, the fearch of truths that are advantageous to the human fpecies.

CHAPTER II.

Whether the Volatile Alkali is a certain Remedy against the Bite of the Viper.

I DEEMED it neceffary to examine this firft queftion in the moft circumftantial way, and therefore
multiplied the experiments extremely, and diverfified them very much. This is the only method
that could lead to demonftration, and 1 flatter myfelf that my readers will be freed from all doubt.

The animals I had bit by vipers were of three different kinds. I employed birds and quadrupeds
with warm blood; and frogs, which have the
blood cold.

Amongft birds, I almoft always employed fparrows, pigeons, and fowls; amongft quadrupeds,
rabbits, guinea-pigs, cats, and dogs.

An animal may be bit by a fingle viper, and by
feveral. It may be bit once or more; in a fingle
part, or in feveral.—All thefe cafes may make great
variations in the difeafe and effects of the venom;
it was therefore neceffary to diftinguifh them from
each other.

K 2 *Animals*

Animals bit by a single Viper, and only once.

The leg is the part of the animal I conftantly had bit by the viper, in all the experiments contained in this chapter. By *leg*, I mean the mufcular part of the foot, that is betwixt the femur and tarfus. The facility of having animals bit in this part by the viper, made me give it the preference. There is likewife another advantage, the eafe with which the remedies are in this cafe applied.

In the experiments of this chapter, as well as thofe of the following one, I employed no other remedy againft the bite of the viper than the *fluid volatile alkali*, to be found in every apothecary's fhop. Some that I made ufe of, I made myfelf. Its compofition has been long known, and is defcribed in all the pharmacopeias. I employed it by having it fwallowed, and by applying it to the part. When I wifhed to treat the part bitten, I dapped it a long time with a piece of linen, well moiftened with the volatile alkali, and laftly, covered it with the fame linen, to keep it wet the longer. That which was fwallowed, as will be feen hereafter, was diluted with a quantity of water. On many occafions, I repeated it feveral times, and likewife made frefh applications of it to the part. There are animals which live fo fhort a time after they have been bit, that I thought it fuperfluous to make repeated applications of the volatile

latile alkali to the bite. When I fay, fimply, that
I treated the part bit, or that I treated the animal,
it muſt be underſtood, that the volatile alkali was
not given internally, but only applied to the part.

I had a dozen fparrows bit by as many vipers, a
fingle time each. I took them from the cage, one
after the other, without any choice. The firſt that
was bit was immediately treated ; the fecond was
not.; the third was treated, and the fourth not ;
and fo on as to the others, each having a thread
tied to its foot, with knots, to diſtinguiſh them from
each other. The feathers had been previouſly cut
from the legs with fciffars. The animal was fcarcely bit by the viper, when it was treated ; fo that there
was an interval of not more than five or fix feconds
betwixt the bite and the application of the *volatile
alkali*.

The fparrow firſt bit, which was treated, at the
end of two minutes could no longer fupport itfelf
on its feet, and died at the end of fifteen.

The fecond, not treated, began to reel after three
minutes, and died at the end of the thirty-fifth.

The third fell on its belly after fix minutes, and
died at the end of thirty-eight.

The fourth fell after four minutes, and died at
the end of twenty.

The fifth after five minutes, and died at the end
of twenty-feven.

The fixth after feven minutes, and died at the
end of thirty,

K 3   The

The feventh was ftill living at the end of three hours, and did not appear to have at all fuffered.

The eighth fell after two minutes, and died at the end of feven.

The ninth fell after three minutes, and died at the end of eleven.

The tenth fell after two minutes, and died at the end of fifteen.

The eleventh fell after a minute and one-third, and died at the end of two and an half.

The twelfth fell after fix minutes, and died at the end of thirty-two.

The fparrow that was bit the feventh, as I have juft faid, was ftill living at the end of three hours. I examined its leg, and found it perfectly in its natural ftate, without lividnefs, without fwelling, or any apparent wound. The legs of the other fparrows were very much changed, even immediately after they had been bit; whence it was eafy to conjecture, that either the fparrow above alluded to had not been bit by the viper, or that this laft was without venom.

To difcover which of thefe two hypothefes was the true one, I had this fparrow bit by the fame viper, in the fame leg. A little blood flowed from the wound, which I immediately treated. It fell after two minutes, and died at the end of four; a proof that the viper was provided with venom, but that the leg had not really been entered by the teeth. I had, however, no fufpicion of this at firft, as the creature bit in the ufual way.

I wifhed

I wifhed to repeat the fame experiment on twelve other fparrows, with the fame order and circum-ftances. But I made the fix that were treated like-wife fwallow a few drops of water, in which I had put a proportion of the volatile alkali, of about an hundredth part.

The time of the death of thefe animals is ex-preffed by the following numbers, reprefenting as many minutes elapfed after the bite; 10. 7, 8, 9, 6, 7, 3, 7, 15, 18, 5, 37. The fix firft numbers fhow the time the fparrows lived, that were treated with the volatile alkali.

From the preceding experiments the following confequences may be deduced:

I. That the vipers I employed were fufficiently provided with venom to kill fparrows.

II. That the venom is fcarcely introduced into the leg of the animal, when it fwells in a fenfible de-gree, changes its colour, and becomes fomewhat livid.

III. That it is not fufficient to enable the venom to infinuate itfelf, that the viper feizes an animal be-twixt its teeth, and that it clofes its mouth, and preffes with it.

IV. That the fluid volatile alkali does not pre-ferve the lives of the fparrows bit by the viper.

V. That the volatile alkali given internally to fparrows, may even be hurtful to them. The fpee-dier death of thofe that fwallowed it, may at leaft lead one to fufpect fo.

K 4 But

But the number of the experiments is not yet fufficient to render the confequences I have juft deduced certain; 'tis a multiplicity of them alone that can effect this.

I had twelve other fparrows, equally lively, bit in the leg as above, each by a fingle viper, and only once. I treated only fix of them with the volatile alkali. They all died. In all of them the leg that was bit became livid, and fwelled in a greater or lefs degree, in lefs than two minutes.

The fix treated died in 3, 4, 6, 11, 30, 33, minutes. The fix that were not treated in 4, 4, 7, 11, 18, 35.

To obtain ftill more certain confequences, I had twenty-four others bit. I treated twelve, and made them fwallow the volatile alkali. All the twenty-four died. The following numbers fhow the minutes the twelve that were treated lived, 2, 3, 3, 5, 5, 5, 7, 7, 10, 15, 15, 22; and thefe again indicate the minutes that thofe furvived on which no remedies were tried, 4, 6, 6, 6, 7, 7, 9, 9, 9, 10, 15, 20.

It is a truth then, eftablifhed by experiments, that the fluid volatile alkali is altogether ufelefs, whether it is fimply applied to the part bit by the viper, or fwallowed by the animal at the fame time. We may even fufpect it to be hurtful, to fparrows at leaft.

However evident it may appear, that the volatile alkali is not an efficacious remedy in this cafe to a fmall animal like a fparrow, it is not on that account

demonftrated,

demonſtrated, that it may not be ſo to a much larger animal, and of a different ſpecies.

The venom introduced into the body of a larger animal, ſhould be conſidered as diminiſhed in quantity. Its effects ſhould certainly not be ſo violent; and indeed this is the caſe with all the poiſons that we know of. What is a remedy to a large animal, or one of a full ſize, may be a poiſon to a ſmaller animal, or to one ſtill young.

We muſt therefore again have recourſe to experiment, and ſee the effect the bite of the viper has on other animals.

Experiments on Pigeons.

I had a pigeon bit in the leg by a viper, and inſtantly treated the part. At the end of a minute it fell forward, and could no longer ſupport itſelf. In twenty ſeconds more it died.

I had another pigeon like the firſt bit in the ſame way, but did not treat it. At the end of two minutes it fell forward, and in two minutes more it died.

I had two other pigeons bit in the leg; one was treated, and the other not. The firſt fell at the end of three minutes, and died at the end of the twentieth. The other fell at the end of a ſingle minute, and died likewiſe after the twentieth.

Of two other pigeons bit in the leg, I treated only one. The one treated died at the end of forty hours, the other at the end of an hour.

I had

I had fix other pigeons bit in the ufual way. Three were treated, and three not. Thofe that were treated died at the end of 6, 22, 40, hours. The other three died at the end of 1, 2, 10, hours.

I had two others bit in the leg in the ufual way; one I treated, the other I did not. The treated one died at the end of eight minutes; the other at the end of two hours.

The intervals at which pigeons die that are bit by the viper are fo different, that they fcarcely allow a reafonable conjecture. It feems, however, that two truths may already be deduced. One, that the *volatile alkali* does not preferve from death the pigeons bit by the viper. The other, that birds larger than fparrows live longer in the fame circumftances; or, if you will, that pigeons die much later than fparrows.

But experiments muft be multiplied, and the circumftances attending them examined more attentively.

I do not conceive very well how it was, that of two animals of the fame kind, bit once in the fame part, one died at the end of two minutes, and the other at the end of 40 hours.

I likewife obferved fomething fimilar to this in the fparrows; and therefore determined at length to have a very large number of both kinds bit. I did not treat any of them; but, in return, I marked carefully all the circumftances that attended the experiments. I fhall not enter into a detail of them here, on account of the very great number of them;

but

but think it fufficient to deduce the following truths:

I. That other circumftances alike, the larger the viper, the more violent the difeafe, and the more fpeedy the death.

II. That the difeafe increafes likewife in violence, in proportion as the viper is more enraged.

III. That it likewife augments in proportion to the time the viper compreffes the animal it has bit betwixt its teeth.

IV. That the difeafe of the part bitten feems to be greater in proportion to the time the animal fur-vives.

V. That in fome animals black and livid blood flows from the wound, as foon as it is made.

VI. That in others, on the contrary, it flows red, and preferves that colour.

VII. That the animals from which the red blood flows, die later than thofe from which it flows black and livid.

VIII. That the venom likewife, which preferves its colour and its qualities, fometimes flows out with the blood. In which cafe, the animal not on-ly furvives, or is much longer in dying, but fome-times does not appear to have had any complaint.

Thefe confequences, the fruit of an infinite num-ber of experiments, diverfified in every poffible way, and in which all the circumftances that accompanied them were rigoroufly examined, form fo many principles, which explain how it is that of two ani-mals

mals bit in the fame part, one dies fuddenly, and the other furvives, or does not die till very late.

There is likewife another reafon, which I have fince difcovered, and which may vary the effects, in animals that have been bit, very much. This is owing to the viper itfelf. I have fometimes, tho' very rarely, found vipers that had no venom in either of the two veficles, and more frequently, that only had it in one.

What led me at firft to fufpect that the veficles did not conftantly contain venom, was obferving it to be to no purpofe that I had a pigeon bit repeatedly by a certain viper; and that it not only did not die, but difcovered no fymptom of difeafe, notwithftanding the canine teeth of the viper had pierced its flefh through in feveral places.

Having had occafion, in the courfe of thefe experiments, to cut of the heads of a great number of vipers, and to examine their venom, out of two hundred, perhaps, I found two that were entirely deftitute of venom, and five that, inftead of venom, had a white and opake vifcous matter in the veficles. In two of thefe laft, I found this white matter to be perfectly innocent. But in the other three it ftill preferved, partly at leaft, its venomous quality, as I affured myfelf by introducing a fmall quantity of it into the legs of pigeons, which had been bit fuperficially, and which died at the end of a few minutes.

It is another eftablifhed truth then, that vipers are fometimes found without any venom, and that

fomewhat

fomewhat more frequently a whitifh humour is contained in their veficles, which is not always venomous. But thefe cafes are invariably very rare, and
only met with in examining a very great number of
vipers; whence it follows that it is alfo true, that
vipers have in general their veficles filled with venom, and that this humour occafions difeafes, and
even death.

I obtained much more uniform eonfequences, by
introducing the venom into the body of the animal,
inftead of having it bit by the viper. This is the
method I employed. I cut off the head of a viper
with a pair of fciffars, and, after a quarter of an
hour, opened the mouth, and with another pair of
fciffars feparated the lower jaw. I then divided the
upper part of the head in two with very ftrong fciffars; each part being furnifhed with the canine
teeth, and with the veficle of venom. With a little
courage and dexterity, which are acquired by cuftom, it is eafy to force the tooth of the viper, on
which a compreffion is made with the fore finger
whilft the veficle is preffed upon by the thumb, into
the fkin of an animal. A greater or lefs quantity
of the venom may be introduced, by preffing more
or lefs on the veficle; the wound may be made
wherever one pleafes; and, laftly, the venom may
be kept from being rejected, by letting the tooth
remain a long time in the wound. A great number
of experiments made in this way, prove that fparrows
die betwixt five and eight minutes, and pigeons in
betwixt eight and twelve. There are very few that

die

die fooner or later ; whence it follows, that by purfuing this method, the periods of their difeafe are
both fhorter and more uniform.

I had a dozen pigeons bit in the ufual way, one
after another, by as many vipers, and treated them
all with the volatile alkali. They all died. The
numbers 4, 10, 16, 52, exprefs the time in minutes
in which four of thefe pigeons died; and the numbers 2, 4, 9, 15, 19, 22, 25, 36, exprefs the time,
in hours, of the death of the others.

Thefe new experiments leave no doubt as to the
inefficacy of the fluid volatile alkali againft the venom of the viper.

To affure myfelf ftill better of this, I had twenty-
four other pigeons bit, each of them once in the leg,
by a viper. I treated them all, but only twenty-
two died. The time of their death is expreffed in
minutes, by the numbers 4, 4, 6, 6, 7, 8, 8, 10, 12,
14, 14, 20, 50, 50, 56; and in hours, by 1, 1, 2,
4, 7, 10, 18, 26, 30.

Two of thefe pigeons, bit in the fame way as the
others, appeared not to have fuffered at all, walking
about the chamber as before the operation. At the
end of two hours, being defirous of examining the
ftate of their legs, I could find no fign of difeafe.
They were neither fwelled nor livid. I could only
find in one of them a fmall hole, and a fmall red
fpot of blood, at the part where the tooth had penetrated. Since there was not the fmalleft mark of
difeafe, it was eafy to perceive that the venom had
not introduced itfelf ; or, if it had, that it had been

2   thrown

thrown out again, so as not to occasion any complaint to the animal. After ten other hours, I had both pigeons bit once in the same leg by two vipers that had already been employed in the same way. At the end of three minutes there were signs of disease : one died at the end of an hour, the other at the end of two.

Not content with these experiments, I had twelve other pigeons bit in the usual way. I treated them immediately, and made them swallow the volatile alkali. They all twelve died, at the end of 4, 4, 7, 10, 10, 10, 15, 18, 20, minutes ; and of 2, 3, 3, hours.

Whilst it is certain, on one hand, that the volatile alkali is of no effect in recovering pigeons bit by the viper ; on the other hand it remains undecided whether it is in this case hurtful or not.

The periods at which these animals die are so various, that it is not possible to deduce any certain consequences from them.

Experiments on Fowls.

It is not sufficient to have demonstrated the inutility of the fluid volatile alkali administered to pigeons, to allow us to conclude that it is useless to larger animals, that are more difficult to kill. The volatile alkali may have time to act against the venom of the viper, when the disease is not so violent, and the animal slower in dying.

There are certain remedies which, although efficacious,

cacious, require a certain time to act; and, indeed, almoſt all are of this deſcription.

I had a fowl bit once in the leg by a viper, and immediately treated it; at the end of ſix hours the fowl died. I afterwards had another bit once by a viper, and did not treat it. This one died in eight hours.

I had two other fowls bit once in the leg as uſual. One was treated; the other not. The firſt died in four hours; the other in ten.

I had ſix other fowls bit as above, each once in the leg by a diſtinct viper. The three firſt were treated with the volatile alkali, and died; one in ſix hours, another in eight, and the third in nine. The three others were not treated, and died in 7, 9, 20, hours.

Although the number of the experiments hither-to made on fowls, is not yet ſufficient to allow cer-tain conſequences to be drawn, it however appears, that the following very probable ones may already be ſtated.

I. That it is very poſſible for fowls bit once in the leg by a viper, to die.

II. That they in general die much later than pi-geons; and than ſparrows, which die again with much greater facility than pigeons.

III. That birds reſiſt death in proportion to their ſize.

IV. That the volatile alkali is not only of no uſe in curing fowls bit by the viper, but that it is pro-bably even hurtful to them.

But

But experiments muft be multiplied much more, to fee if the confequences juft deduced are well or badly founded.

I had therefore fix fowls bit feparately by fix vipers, each once in the leg. I treated them all fix, and made a frefh application of the volatile alkali to the part bitten, every two hours. Two of the fowls died in the fpace of four hours, one in five, two in fix, and one in ten. A moment after, I had fix other fowls bit by as many vipers, each once in the leg, and did not treat either of them. Two died in two hours, two in ten, and two in twelve.

Twelve other fowls were bit by as many vipers, each once in the leg. I treated fix, and made them fwallow the volatile alkali. The other fix were left to themfelves. Of the fix treated, five died; the fixth had fcarcely any fymptom of complaint. Its leg neither fwelled, nor became at all livid. There was fimply a hole in the fkin, which was red and a good deal inflamed. The five I have juft mentioned died in 3, 4, 6, 7, 10, hours. The other fix died in 6, 10, 17, 22, 36, 36, hours.

Had the experiments I have related fo far been more numerous, the abfolute inutility of the fluid volatile alkali againft the bite of the viper would not only have been demonftrated, but we might even have doubted its innocence, at leaft to this fpecies of animals.

The treated fowl that did not die, proves nothing in favour of the volatile alkali, as will be feen in the continuation of this work. It is one of the cafes I remarked above, in fpeaking of the pigeons and fparrows, in which the venom was not communicated to the part bit, although the canine tooth had left fome opening in it, either owing to the viper not having any venom, or to the rejection of it. Nothing is found in either of thefe cafes to favour the volatile alkali.

Having affured myfelf of the inutility of this remedy to the three fpecies of birds I have fubmitted to the experiment, I think it time to make the fame trials on quadrupeds.

Experiments on Guineapigs.

I had a large guineapig bit once in the leg by a viper, and immediately treated it. In a little time the leg fwelled, and became livid. At the end of fixteen hours a wound of an inch in breadth formed itfelf at the part that had been bit and treated. In twenty hours the fkin in this part was entirely eaten away. The wound continued open for more than twenty days, during which time the animal moved its leg with difficulty; the foot was greatly contracted, and the mufcles very much difeafed. The animal recovered however, but its leg ftill remained in a degree contracted, and it could never recover the perfect ufe of it.

Another guineapig, almoft as large as the former, was, in the fame way, bit once by a viper in the

leg,

leg, which was not treated. The animal died at
the end of two hours.

I had four others, of fcarcely a third the fize of
two preceding ones, bit in the above way. I
treated each of them, and made them fwallow the
volatile alkali. They all died, one in two hours,
another in three, the third in fix, and the fourth
not till the twentieth hour had elapfed.

That I might have a comparative experiment, I
had four other guineapigs, entirely like the preceding ones, bit, and did not treat either of them.
They all four died, one at the end of feven hours,
another at the end of ten, the third at the end of
thirty, and the fourth at the end of thirty-one.

From thefe experiments we may, I think, already draw the following inferences, which if not
certain, are at leaft very probable.

I. That the bite of the viper is capable of killing even the largeft guineapigs.

II. That the fmaller animals die fooner than the
larger ones of the fame fpecies.

III. That the volatile alkali is not a certain remedy againft the bite of the viper.

It may be objected, that the firft guineapig bit
and treated, at length recovered, and that all which
were not treated died. This is true, but proves
nothing, fince, as has been feen above, there are
feveral circumftances that may render the bite of
the viper innocent; and, on the other hand, we
have feen that the other four guineapigs died, although they were treated. Now if we confider

L 2   that

that the four treated died in a much fmaller fpace of time than the five that were not treated, we may fufpect that the volatile alkali was more than ufelefs, that it was hurtful.

To remove all doubt, I had twelve other guinea-pigs bit, all alike in fize, and fimilar to the eight preceding ones. Six were treated, and fix not.

The firft I had bit was the fame I have fpoken of a little above, and which, far from dying of the bite, had not even the difeafe of the venom. Although treated, it died at the end of thirty hours. The five others that were likewife treated, had all of them the difeafe of the venom, but only three died; two in lefs than twenty hours, the other at the end of twenty-feven. The two that furvived had each of them a large wound in the leg that had been bit, and this remained open for more than ten days.

Of the fix that were not treated two only died, in lefs than fixteen hours. Three others had deep wounds, which remained open for feven days, and then healed. The fixth had not the fmalleft fymptom of difeafe, and I could not difcover in its leg any mark of the viper's tooth having penetrated.

The cafes fo far related, feem to leave no doubt as to the inutility of the volatile alkali, when tried likewife on thefe animals; and they do not remove the fufpicion, that it may poffibly be even hurtful to them.

We likewife fee that the fmaller and younger guineapigs die fooner than the larger ones.

I had

I had a doz— ..y ſmall ones bit, each ſcarcely weighing five ounces. Six were treated, and ſix not. Thoſe that were treated died in 30, 40, 50, minutes, and 1, 2, 3, hours. Thoſe that were not treated in 57 minutes, and 2, 3, 4, 4, 4, hours.

I afterwards had ſix guineapigs bit, three of the largeſt of which were treated ; the other three were not. Of thoſe that were treated only one died, and only one again of thoſe that were not treated. All of them, however, were very much diſeaſed, and thoſe that were treated were the laſt to recover.

Experiments on Rabbits.

It remained for me to make the ſame experiments on rabbits, in purſuance of the plan I had propoſed to myſelf.

With this view, I had a large rabbit bit by a viper once in the leg, which I immediately treated with the volatile alkali, making the animal ſwallow the ſame diluted with water. At the end of an hour I repeated the application and the internal remedy. The rabbit died at the end of three hours, with very ſlight marks of diſeaſe in its leg.

I had another, perfectly like the former, bit in the ſame way once in the leg by a viper, and at the ſame time. It had ſlight ſymptoms of the diſeaſe of the venom, and the leg became ſomewhat ſwelled. At the end of thirty hours a wound two lines in breadth, and very deep, appeared on the ſkin at

the

the part where it had been bit. ᴀᴛ.. five days more the animal was perfectly recovered.

The refult of two experiments alone can be in no way certain, I therefore had recourfe to my ufual method.

I had a dozen rabbits of a middle fize bit by as many vipers, each once in the leg. Six were treated, and fix not. Only two died of thofe that were treated, and three of thofe that were not. Two of the four treated ones that did not die fcarcely had any complaint. Their legs were but little fwelled, and were not livid. The other two were very much difeafed, and had large wounds that were four days in healing. Of the two that died, one lived two hours, the other five. The fix that were not treated were all of them very much difeafed, and had large wounds in their legs, which fwelled violently, and became very livid. The three that died lived 14, 22, 47, hours; the others did not recover till the end of the feventh day.

It is a conftant obfervation, that when the animal bit by the viper dies very foon, the bitten part is proportionably lefs changed, lefs fwelled, and lefs livid. The change which takes place at the part where the poifon has entered, I call the external difeafe, to diftinguifh it from the other, which is infinitely more violent and dangerous, and which kills the animal in a more direct way. I fhall fpeak more fully of this laft in the fourth chapter of this fecond part, in which I fhall endeavour to account for this particular.

The

The few experiments hitherto made on rabbits may already make us fufpect the little efficacy of the volatile alkali, which we may be even tempted to believe hurtful. It is certain in the interim, that middle-fized rabbits frequently refift the venom of the viper.

I wifhed to try the effects of this on much fmaller ones, and for this purpofe had a dozen bit in the ufual way. I treated fix, and did not treat the others. All the twelve died; the treated at the end of 2, 3, 4, 6, 8, 9, hours; and the others at the end of 3, 5, 7, 9, 12, 18.

I repeated thefe experiments on twelve other fmall rabbits, exactly like the foregoing ones. I treated fix, and made them fwallow the volatile alkali every hour. The others I did not treat. They all died; thofe that were not treated, at the end of 1, 1, 2, 2, 5, 17, hours; the others in the fpace of 1, 3, 3, 10, 16, 16, hours.

Thefe new experiments already fhow very clearly the little efficacy of the volatile alkali againft the bite of the viper, when tried on rabbits; they even lead me to fufpect it to be rather hurtful than otherwife.

We likewife fee that fmall rabbits die from the bite of the viper, whether they are treated or not; but that the larger ones frequently furvive its effects.

In confequence of this, I had fix of thefe animals, very large ones, bit each by a viper once in the leg. Three were treated, and fwallowed the vola-

tile

tile alkali. Two of thefe died at the end of twenty hours; the third was very much difeafed, and had a large wound, which remained open for twenty-three days. Of thofe that were not treated, one died at the end of thirty-four hours; the other two had the difeafe of the venom, but recovered in lefs than ten days.

I repeated this experiment in the fame way on fix other large rabbits. Of the three that were treated, one died; and one likewife died of the three that were not treated. The other two of thefe laft recovered in ten days; and the two treated ones that furvived, not till the end of eighteen.

I think there can be no longer any doubt of the inefficacy of the volatile alkali to thefe animals; on the other hand, inftead of diminifhing it, it feems to ftrengthen and reinforce the difeafe.

It remains to try the effects of the bite of the viper on cats and dogs. The number of my experiments on the animals of thefe two fpecies is much fmaller than that of thofe on the others. The difficulty of procuring them, the danger one runs in operating on them, and, ftill more, the inconvenience of keeping them during the long continuance of the difeafe, and the unpleafantnefs of feeing them fuffer, have occafioned me to do lefs in this inftance than the fubject may perhaps appear to have required,

Experiments on Cats.

I had two very fmall kittens bit in the ufual way, each once in the leg. One was treated, the other not. The laft died at the end of fixteen hours. The treated one was exceedingly ill, and had a wound which remained open for five days, in its foot. It lived, however.

Three very fmall kittens were brought to me, ftill younger than the foregoing ones. I had them bit, as ufual, in the leg. I treated one, and made it fwallow the volatile alkali. I did nothing to either of the other two. They all three died in lefs than fix hours.

Thefe experiments are neither fufficiently uniform, nor enough in number, to admit the drawing of certain confequences from them. We fee that, in general, the fmaller animals of the fame fpecies, fuch as cats, for inftance, die much readier than the larger ones ; and likewife that thofe die which have been treated, and have fwallowed the volatile alkali.

I had two other kittens bit, larger than thofe I employed before. Each of them was, as ufual, bit once by a viper, One was treated, the other was not. Neither of them either died, nor was very ill. They had no wound ; and both of them at the end of twenty-four hours ate very heartily. The leg, however, in each, was not yet very fupple in its motions. I did not make the one I treated fwallow the volatile alkali, on account of the difficulty

one

one meets with in attempting this, when these
creatures are pretty large.   They become extremely furious, and are very difficult to manage, at least
without risk.

I had two other kittens of the same size of the
preceding ones bit, and treated neither of them.
They were each of them bit once in the leg.   They
both recovered, and had no perceptible wound.
It was twenty hours indeed before they had any use
of the leg that had been bit; however, they seemed perfectly recovered at the end of the third day.

Two large grown cats were, in the same way, bit
in the leg.   Neither of them was treated, and neither died.   At the end of sixteen hours they fed a
little, and could already use their legs, although not
very well.   At the end of thirty hours they appeared to be perfectly recovered.

Scarcely has a cat been bit in the leg by a viper,
when it can no longer make any use of the part.
It lies down, and continues longer in this posture
in proportion to the violence of the disease.   It
neither eats, nor drinks till the symptoms abate,
and when that happens, recovers to a certainty.

Experiments on Dogs.

We are now to try the effects of the volatile alkali, which has been of no utility to the cats, on
dogs that have been bit by the viper.   The dog has
a great affinity to man himself, and is, of all animals, the one the most susceptible of the passions.

It

It is certainly much more fo than the cat and the other animals, that have been bit in the courfe of thefe experiments. Dogs are to. be met with of every fize, even fo large as not to differ much, in that refpect, from an adult perfon.

The effects of the bite of the viper on dogs, may be of great ufe in judging of the bite of the viper in man himfelf.

I had two dogs of a middle fize bit once in the leg. I treated one of them every two hours, and made it fwallow the volatile alkali as often. Neither of them died, although the leg was fwelled in each. The one not treated had no wound, and recovered at the end of four days; the one that was treated had a large wound, and did not recover till the clofe of the tenth day.

I had two other much fmaller dogs bit, and treated only one of them. They both died in lefs than three hours, with a degree of fwelling and lividity in the part bitten.

Two large dogs were brought to me, and I conceived from their fize, that they would recover although not treated. I had them bit in the ufual way, once in the leg. One fcarcely had any fenfible complaint; the other no perceptible wound. The leg of the laft, however, fwelled very much, and did not get well till the end of the fixth day.

I had two other large dogs bit by a viper as ufual, each once in the leg, and did not treat them. One recovered in two days, the other in fix.

From

From the experiments hitherto made on dogs,
we may draw thefe conclufions :

I. That the fmaller ones ufually die from the
bite of the viper.

II. That large ones generally recover.

III. That of the middle fized ones, fome recover,
and fome die.

IV. That the volatile alkali feems to be neither a
certain nor a ufeful remedy againft the bite of the
viper,

Experiments on Frogs.

It remained for me to try the effects of the venom
of the viper on frogs. I had hitherto operated on
animals with warm blood ; it was likewife neceffary
to make fome experiments on thofe that have the
blood cold.

I had a dozen frogs bit by as many vipers, each
once in the leg. I treated fix of them only. Two
of thefe died at the end of twenty hours, and the
legs of the other four fwelled, and were a little livid ; they recovered however. Of the fix not
treated, three died at the end of five hours. Of the
three that furvived, one had a fwelling and difco-
louration of its leg ; the other two had no apparent
complaint.

The confequences were as yet too vague, and too
few in number, to admit any certain conclufions to
be drawn from them,

I there-

I therefore had a dozen others bit in the fame way, and treated fix of them only. To thefe I renewed the application of the volatile alkali every hour, making them fwallow it at the fame time. All the fix, one of which did not furvive the twentieth minute, died in lefs than four hours. Of the fix not treated, four died at the end of 6, 10, 12, 20, hours; the fifth had fcarcely any complaint, and the fixth recovered two days after.

I repeated this experiment on twelve other frogs, having them bit in the fame way, each once in the leg by a viper. Six were treated every hour, and fwallowed the volatile alkali, as often. The other fix were left to themfelves. Five of the firft died; the fixth had fcarcely any fymptom of complaint. Of the fix not treated, three died, and the other three recovered at the end of two days.

After what has been faid, I think there can no longer be any doubt of the inutility of the fluid volatile alkali. It is very probable that, when given internally to frogs, it increafes the difeafe caufed by the venom, inftead of diminifhing it. It is at leaft certain, that the animal dies the readier under thefe circumftances.

C H A P T E R   III.

Of the Effects of the Bite of one or several Vipers, on the same Part of an Animal, or on two corresponding Parts of the same Animal.

I HAVE hitherto fpoken of the effects of the venom on animals bit once by a viper, in the fame part. It now remains to fpeak of animals bit repeatedly by one or more vipers in different parts.

It is natural to conceive, that a viper which bites the fame animal feveral times, muft bring on a difeafe proportionably violent. After having feen in the firft part of this work, that the venom of the viper is a humour feparated from the fluids of the animal, and fecreted in a veficle or gland ; and that this humour is always venomous in itfelf when it is introduced by a wound into the bodies of animals, particularly of thofe with warm blood ; there can no longer be any doubt of this truth, nor of the abfolute falfehood of the hypothefis of Monf. Charas, who pretends that the venom of the viper is entirely occafioned by the fury of the animal, which changes the faliva and other humours of its mouth, to fuch a degree as to produce a powerful venom, fuch as is obferved in the foam of a mad dog.

The

The veficle is moreover conftructed in fuch a way, that all the venom cannot flow out at once, at a fingle bite, however forcible it may be, and however the viper may be enraged. A defcription of this veficle, with that of the gland, will be feen in the third part of this work. From the foregoing confideration it was neceffary to examine the effects and difeafes produced by feveral bites, although of a fingle viper. There are feveral examples of perfons bit more than once by the fame viper; and notwithftanding this cafe is not one of the moft frequent, it occurs however from time to time.

It is not only very important to examine the effects of the repeated bites of the fame viper on the fame part of an animal; but likewife to obferve the action of the venom on the different parts of the fame animal.

We know that an animal is formed of organs and parts, differently organized. There are parts that have veffels and nerves, without having mufcles; and thefe are in different proportions, and differently diftributed: there are others again that have no nerves, and, if they have any, have only a few very fine capillary veffels. It is natural to fuppofe that the effects of the venom of the viper, on parts of an animal fo very different from each other, muft be altogether different; and that the fame quantity of venom conveyed into a wound made in an animal, may produce either death, a flight difeafe, or none at all. In a word, it appeared to me, that nothing ought to be omitted in fo important a matter.

3

There

There is likewife a cafe, although I think it a very rare one, in which feveral vipers together bite the fame part, or different parts, of an aniinal. However rare this accident may be, it is not impoffible for it to happen; and it is not an extraordinary thing to find, at certain times of the year, feveral vipers collected together. A man who may not have noticed this, may by treading upon them be in danger of being bit by more than one; and I knew a viper-catcher who was bit in the hand by two at the fame time, and who might have been bit by more than two, fince feveral of them were making their way at the fame time out of a box.

Thefe examples of animals bit by feveral vipers may, however, agree very well, making fome little allowance, with the cafes of the repeated bites of the fame viper, whether on the fame part, or on different parts, of an animal.

I faid above, that I had found by experience the effects of the venom to be much more uniform, when, inftead of having the animals bit by vipers, the venom is conveyed into them, by preffing with one finger the veficle which contains it, whilft with the other the tooth of the viper is forced into the part. I have frequently employed this method during the courfe of my experiments, particularly in thofe on the fparrows and pigeons. In this way I not only fucceeded in wounding the fame part of the animal over again to a certainty, but even the very fibre. I could likewife affure myfelf, if I was defirous of it, whether the veficle contained venom, or

whether

whether the quality of the latter was fufpicious or changed.

The flighteft compreffion made on the veficle is fufficient to bring a very fmall drop of venom to the point of the tooth; its tranfparent colour determines its activity and nature.

The firft thing I thought it neceffary to determine here, was to fee whether the fecond bite of the viper is as powerful as the firft, the third as the fecond, and fo on as to the others; and how many times, one after the other, the viper can venom animals with its bite. I took a viper of a middle fize, and very lively, and, without provoking it much, made it bite a pigeon once in the leg. The pigeon died at the end of twelve minutes. A moment after it had bit this one, I made it bite a fecond, a third, a fourth, a fifth, a fixth, and a feventh, in the fame part. The fecond died at the end of eighteen minutes, the third of fixteen, the fourth of fifty-two, and the fifth at the end of twenty hours; the fixth had fcarcely any fymptoms of complaint, and the feventh continued perfectly well.

I repeated this experiment feveral times, and the confequences were fomewhat various. I met with fome vipers, particularly the largeft of them, that could kill ten, and even twelve pigeons. If they are very much enraged during the firft bites, the laft, as I have affured myfelf by repeated experiments, are lefs dangerous.

It is an eftablifhed truth then, as I have feveral times experienced, that the firft repeated bites of a

viper

viper are almost equally dangerous; and that in proportion as a viper is enraged, the difeafe occafioned by its bite is more violent.

This laft truth may tend in fome degree, to account for the treacherous experiments of Charas on the venom of the viper. In oppofition to Redi, as has been feen above, he was of opinion that this venom confifts fimply in the rage of the animal, and made a great many experiments to fupport his hypothefis.

It was natural to conceive, that the more a viper is enraged, the greater will be the difeafe produced by it, and *vice verfa*. But to draw a certain inference from this, it was firft neceffary to be affured, whether the degree of the difeafe, or intenfity of the venom, is in proportion to the rage of the animal : a very difficult experiment, and perhaps impoffible to make well ; and which would not probably have been yet fufficient, fince after all this might have been an accidental circumftance, and not the true caufe of what was obferved.

Charas, who was ignorant of the true reafon of the greater intenfenefs of the difeafe in the cafes in which the viper is enraged, was miftaken in his inferences. It is not furprizing that the naturalift fhould here take that for the caufe, which is the effect of the circumftances that accompany it.

There are three reafons why the bite of an enraged viper is more dangerous than that of one which is not enraged. The firft is, that the more a viper is enraged, the deeper it forces its teeth into the ani-

mal; the fecond, that it keeps them there a longer
time ; the third, that without letting go the part it
has bit, it continues to contract the mufcles that
comprefs the veficle of venom.

When one has been fome time accuftomed to
have animals bit by vipers, it is not difficult to perceive the truth of the firft reafon ; and it is fometimes even obferved, that the tooth of the viper
pierces the fkin of the larger kind of quadrupeds
with great difficulty, or only imperfectly and in
part. All my experiments have fhown me, that the
difeafe is in general more violent, in proportion as
the venom has introduced itfelf deeper into the fkin
and other parts of the animal.

The fame obfervation likewife demonftrates the
truth of the fecond reafon. We frequently fee
that when a viper is violently enraged, it does not
eafily let go its hold ; one might even fay that it
finds a difficulty in withdrawing its teeth. In thefe
cafes it is eafy to perceive, that during all this time
the tooth not only prevents the venom from being
thrown out again with the blood that naturally flows
from wounds ; but likewife, that it facilitates its
union and mixture with the fluids of the animal.

The third reafon is ftill of greater weight than
either of the other two. It has been feen, that feveral bites of the viper are neceffary to empty the
veficle of the venom perfectly. It has been feen,
that the firft bites of the viper are nearly of the fame
activity, becaufe they are fucceeded by the flowing

of nearly an equal quantity of venom. The cellular ſtructure of the veſicle does not allow it to be eaſily emptied, nor at once. When the viper keeps an animal a long time compreſſed by its teeth, and is very much enraged, it continues viſibly to contract the muſcles of its jaw. The muſcles which ſurround the veſicle alternately relax and contract without interruption, ſo that in theſe caſes we may reckon the bite of the viper, not as a ſingle one, but as ſeveral ; and this may be carried to ſuch a length, that the viper, almoſt exhauſted of its venom, may not be capable of killing a ſmall animal.

It has been ſeen, that the firſt bites of the viper are all nearly of the ſame activity, and that it is only the laſt which exhibit a marked difference ; and this I have accounted for.

After what has been ſaid, it is natural to conceive that the diſeaſe of the venom muſt be more dangerous, if the viper has bit the ſame animal ſeveral times. I have aſſured myſelf of the truth of this, by experiments the detail of which I ſhall not enter into here, as it would be tedious, and would not beſides anſwer any great purpoſe.

In inveſtigating this ſubject, I took care to employ animals of the ſame ſize and ſpecies, and had them bit by vipers ſimilar to each other. I more commonly availed myſelf of my uſual method, and the conſequences were ſtill more uniform. When only a few experiments are made, the conſequences may be equivocal, ſince it can ſcarcely hap-

pen

pen that the circumſtances will be perfectly the ſame. They may not only differ on account of the quantity of venom that remains in the animal's wound, which is ſubject to greater or leſs variations, but likewiſe becauſe it is very difficult to wound the ſame fibres, and the ſame veſſels, Theſe variations do not fail to occur; but in a great number of experiments, the circumſtances counterbalance each other, and ſo great a variety of conſequences preſents itſelf, that there is not the ſmalleſt danger of being miſled by them. Such has been at leaſt my opinion as to thoſe I have obtained.

A new enquiry to be made, was to know if the diſeaſe would be equal, whether a ſingle part was bit ſeveral times by a viper, or two different parts, provided the number of bites was the ſame.

This enquiry coſt me a vaſt many experiments, which I was obliged to make with the ſame circumſtances, only varying the part bit.

I not only tried it on birds, but on a great number of quadrupeds. I had them bit in the ſame part of their legs. I compared thoſe that were bit in both legs, with thoſe that were only bit in one, the total number of bites being the ſame in each animal.

Here again the conſequences were more or leſs conſtant. I was obliged to multiply my experiments till I conceived that I could advance, with great probability, the two following poſitions.

M 3 I. That

I. That an animal dies fooner when bit a certain number of times in two parts, than when the fame number of bites is confined to one.

II. That in this cafe the fingle part is fubject to a much more violent external difeafe.

By external difeafe, I mean the fwelling of the part that has been bit, the livid and black colour of the fkin and blood, and the wound that forms a fhort time after the bite. Thefe fymptoms are certainly more violent when the part has been bit feveral times; although it is a fact, as will be feen hereafter, that the animals die much later, and that fewer of them die in proportion. It is however to be noticed, that this only happens when the animals live for fome time after being bit, fince otherwife the venom has not an opportunity of effecting any notable change in the external parts ; in fo much, that if the animal dies almoft immediately after being venomed, there are fcarcely any figns of local difeafe.

Before I examine the effects of the bite of the viper on the different parts of an animal, let me be permitted to relate the event of a great many experiments I made on animals of different fpecies, which I had repeatedly bit, and by feveral vipers. In all thefe cafes I employed the fluid volatile alkali, either fimply applied to the part bitten, or given internally at the fame time. Thefe new experiments demonftrate ftill more the inefficacy of the volatile alkali, and how little dependence ought to be placed upon it.

I had

I had fix fowls bit, each of them twice by a diftinct viper. . Three were fimply treated, three were
not. The three that were treated died at the end of
3, 5, 6, hours; the other three at the end of 3, 9, 12,
hours.

I had fix other fowls bit, each by two diftinct
vipers, once in each leg. I treated them, and made
them fwallow the volatile alkali. They were all
dead before the expiration of feven hours; one of
them died in lefs than twenty-feven minutes.

Twelve other fowls were bit, each of them twice
in the leg, and by different vipers. Six only were
treated, and fwallowed the volatile alkali. Nine died
in the whole; five of thofe that were treated, and
four of the others. Two of thefe laft lived forty-
three hours; the treated five that died did not furvive the feventh.

The refult of the laft clafs of experiments, although it does not agree with that of the two that
precede it, is neverthelefs given with precifion. This
fhows how much experiments of this kind may
differ from each other, from circumftances which
occafionally vary, and which cannot always be af-
certained. Thofe that are capable of influencing
the moft are, that vipers are not always provided
with the fame quantity of venom, and that they
are more or lefs vigorous in biting, and in forcing
this humour from the veficle: to thefe may be
added, the effect of a milder or feverer feafon. I
began my experiments in September, and continued
them with more or lefs earneftnefs till the clofe of

the

the January following. I likewise made a few in February, March, and April, and found a sensible difference at these different times. During the severe frost, the vipers were so weak, that it was with difficulty I could get them to bite; and their bites were in a very small degree dangerous.

I cannot here pass over an experiment I made in the month of January, and which at first made me suspect that the volatile alkali might sometimes be a remedy against the bite of the viper.

I had six fowls bit in the leg, each by three vipers, all of which bit three times successively. I treated them several times, and made them as repeatedly swallow the volatile alkali. They all had the disease of the venom, but in a very slight degree, and recovered in a few days.

There remained, as chance would have it, in the same box, eighteen other vipers, perfectly like the eighteen employed in the preceding experiment. Perceiving at the end of fourteen hours that neither of the fowls was dead, and that they were all but slightly diseased, I had six others bit in the same way, each by three of these vipers, and each viper biting three times. I treated neither of them, and only one died, at the end of six days. Two had scarcely any complaint, and the three others recovered on the third day. This experiment demonstrates clearly, that the six treated fowls were not cured by the volatile alkali, but that their recovery was owing to the little vigour and activity of the vipers by which they were bit.

The

The fowl in the laft experiment that was not treated and died, argues nothing in favour of the volatile alkali, fince it is only one out of fix, and fince it did not die till the end of the fixth day. This evidently proves, that if the venom had been in a fomewhat fmaller quantity the fowl would not have died. We have feen above, that a thoufand accidents may vary this greater or lefs quantity of venom, both in the viper that inflicts the bite, and in the animal that receives it.

On this very account, I have made it a maxim, in almoft the whole courfe of this work, to form comparative experiments, and only to compare thofe with each other, that were made at the fame time and with the fame circumftances.

I muft here inform my readers of what befel the vipers I employed laft. The feafon was very cold, and notwithftanding the temperature of my chamber was twelve degrees above the freezing point, the vipers were very fluggifh and inactive. I conceived that I could give them a frefh vigour by additional warmth, and therefore, after keeping them in my laboratory for upwards of fix hours, in a box pierced with holes, I at length placed the box on a fand heat, the warmth of the fuperficies of which was only twenty degrees. At the end of two minutes I found every one of the vipers dead. The fame accident happened to me twice befides, in the fame month, and on an occafion fomewhat fimilar.

Expe-

Experiments on Guineapigs bit several Times, and by several Vipers.

I had two very large guineapigs bit repeatedly in the leg by two vipers. One was treated, the other was not. They both died; the firft at the end of two days, the fecond of thirty-two hours.

I had four other guineapigs, precifely of the fame fize, bit each in the leg by three vipers, three times by each. Two were treated, and fwallowed the volatile alkali; the other two were left to themfelves. All four died in lefs than two days.

Again, I had four others of the fame fize bit in the fame way. They were not treated. One only died, at the end of the fifth day.

Twelve very fmall ones were bit in the fame way. Six were treated, and fwallowed the volatile alkali; the other fix had nothing done to them. They all died in the fpace of twenty minutes.

Two days after, I had twelve others bit, of the fame fize as the laft, each receiving from two diftinct vipers three bites in each leg. Six were treated, and fix not. They all twelve died in two hours. One of the treated ones died in feven minutes, and two of thofe that were not treated in fourteen.

Thefe experiments convince us at a glance of the inutility of the volatile alkali. They likewife fhow, that in animals of this fpecies the fmaller

ones

ones die fooner than the larger, and that their deaths are fpeedier and more certain, in proportion to the greater number of the viper's bites.

Experiments on Rabbits bit feveral Times, and by feveral Vipers.

I had four middle-fized rabbits bit, each four times in the leg, by two diftinct vipers. I treated two of them, which I made fwallow the volatile alkali every two hours, repeating the application as often. They all four died ; the two that were treated, in eighteen hours, the other two at the end of three days. In all of them the difeafe of the venom was very violent, and their legs were very much fwelled.

I had four very large rabbits bit, each by two vipers, twice in the leg. Two were treated, and two not. The two that were treated, although they furvived, continued ill and with open wounds in their legs, for upwards of twenty days. One of the two that were not treated died on the third day; the other recovered on the tenth.

I had a dozen middle-fized rabbits bit in the leg, each by two diftinct vipers, and each viper biting three times. Six were treated, and fix not. Four of the former died, and five of the latter.

Thefe confequences not being either fufficiently uniform, or in a fufficient number, to enable me to

decide

decide as to the volatile alkali. I judged it neceffary to have recourfe to new experiments.

I had twelve rabbits, fomewhat fmaller than
thofe employed in the laft experiment, bit in the
fame way. Six of thefe were treated, and fwallowed the volatile alkali; the other fix were left to
themfelves. All of the former ones died, and five
of the latter; the fixth had fcarcely any perceptible
complaint.

I wifhed to fee whether there would be a fenfible
difference betwixt the effects of the venom, on animals bit a greater or lefs number of times, by a
greater or lefs number of vipers. For this purpofe, I had fix middle-fized rabbits bit, each once
in the leg, by a diftinct viper. I had fix others bit
in the leg each by two diftinct vipers, each of
which made two bites. I had fix others bit in the
fame part, each by two diftinct vipers, each of them
biting four times; and fix others again, each of
which was bit by three diftinct vipers, four times
in the leg by each.

Of the fix of the firft clafs, three died; the other
three had moderate complaints. Of thofe of the
fecond, five died, and the other had a violent attack of the difeafe. All of the third clafs died in
in lefs than forty-three hours; and thofe of the
fourth in lefs than twenty.

Expe-

Experiments on Dogs, bit several Times, and by several Vipers.

I had two small and young dogs bit in the leg, each by two distinct vipers, and twice by each. One was treated, and swallowed the volatile alkali; the other had dothing done to it. They both died in the space of thirteen hours.

I had two dogs, larger by one-half than the preceding ones, bit each by two distinct vipers, and twice by each. One was treated, the other not. Both recovered; the treated one in twenty-six days, the other in ten.

I had four very large ones bit, each by three distinct vipers, three times by each. Two were treated, and two not. One of those that were treated died at the end of the sixth day. The other three were exceedingly ill, and had each of them a large wound in the leg that had been bit.

Two very large dogs were brought to me in excellent order. I had each of them bit in the leg by four well-irritated vipers, each viper biting at least four times. I did not treat them, on account of the difficulty of doing it effectually without the risk of being bit. Both of them recovered in less than ten days. They had wounds, tumour, and lividity, in the part bit. At the end of two days they began to drink, and ate at the end of the third.

Scarcely

Scarcely have animals of any kind, and particularly dogs and cats, been bit by a viper, and are at liberty, when they lie down on the part oppofite to that which has been bit, and in this ftate continue very quiet till they recover. Whenever they begin to drink and to eat, 'tis an almoft certain fign that they will get the better of their complaints. Cats are lefs defirous of food than dogs; I have met with fome that did not eat till after they had been ill feveral days.

That the number of my experiments on dogs might be competent to the purpofe, I procured fix fmall ones, of the fame fize, fpecies, &c. I had them all bit in the leg, each by three vipers, and each viper at three bites. Three were treated, and three not. The three firft died, and only two of the others; the third was exceedingly ill, had a large wound, and did not recover till the end of the fifteenth day.

Not perceiving that the volatile alkali had any good effect againft the bite of the viper, when given to dogs, I thought it proper to purfue my experiments on other kinds of animals.

Experiments on Cats.

This animal makes a very ftrong refiftance to the bite of the viper. This is not becaufe the venom is innocent to it as it is to fome other animals, but becaufe it is very hard to kill.

I had

I had a middle-fized cat bit in the leg by two
vipers, each viper biting twice. I did not treat it.
Its leg fwelled, but not violently. It lay reclined
on its belly during the whole time of its illnefs; it
drank at the end of thirty-fix hours, ate at the end
of fifty-two, and on the fourth day was perfectly
recovered.

I had it bit in another leg by three vipers, each
viper biting twice. Here again I did not treat it.
It vomited feveral times after the fixth hour, and
again after the thirtieth. It drank at the expiration
of forty-two hours, and ate at the clofe of the third
day. On the fifth day it was recovered.

I made choice of another cat of the fame fize as
the former one, and had it bit in the leg by four
vipers, each viper biting four times. I did not
treat it. It fwelled very much, vomited feveral
times, and did not eat till the clofe of the fixth day.

Two days after I had it bit by four frefh
vipers in another leg. It was very much difeafed,
and had frequent vomitings. It ate at the end of
five days, and on the eighth was quite recovered.

I had another cat, larger than the former ones,
and very wild, bit by fix well-enraged vipers, feveral times by each. One of them could not let go
its hold, and was difengaged with fo much difficulty, that its teeth were broken and left in the
flefh. The cat was in a violent rage, but became
tranquil on being fet free. It reclined itfelf on its
belly, as the others had done, vomited from time to
time, and did not eat any thing till after the fifth

3   day.

day. It continued ill two days more, and at length recovered.

It was quite unneceſſary to give the volatile al-kali to the cats, ſince, as we ſee, when they are of a certain ſize, they do not die of the diſeaſe of the venom. Kittens are, however, known to die of it ; and it is likewiſe certain, that grown cats would die too, provided they were bit by a greater number of vipers.

The bite of the viper produces a true diſeaſe in this animal, and this diſeaſe is more violent in pro-portion to the greater number of bites. I cannot, however, preciſely ſay, how many vipers it would require to kill a ſtrong cat of the largeſt ſize. Ten or twelve would, perhaps, be ſcarcely ſufficient.

CHAPTER

C H A P T E R   IV.

Of the Effects of the Bite of the Viper on different Parts of an Animal.

I HAVE hitherto spoken of animals bit by one viper, or by several, either once or repeatedly, but only in a single part; that is to say, in the leg, or in two legs at most. We are now to see the effects of the bite of the viper on the other parts of an animal. It is easy to conceive that the consequences will be somewhat different from those that have already been observed; and that there must be parts in the same animal, more or less susceptible of the venom; several of these, on having them bit, have afforded singular and unforeseen appearances.

Experiments on the Skin.

The part of an animal which is first pierced by the canine tooth of the viper, and which feels before the others the action of the venom, is the skin. I have confined my experiments to the skin of guinea-pigs and rabbits, harmless animals, that are managed without risk. I have not employed birds, as their skin is too delicate for these experiments.

N   Wounds

Wounds made in the skin may be very slight, and altogether external ; they may be more or less deep ; and lastly, they may pierce the skin through and through. I have observed all these cases in the course of my experiments on the bites of the viper. I have sometimes seen the viper's tooth strike the skin so obliquely, that it was either not cut at all, or only superficially. The first case I mentioned, happens frequently, from the viper, when it is enraged, biting at every thing that is presented to it, in any way, and under any form whatever. The second case is much less frequent ; and that in which the bite is made without piercing the skin, still rarer.

These two last cases may happen to man, whose skin may be more or less injured by the canine teeth of the viper.

This research, besides its being curious, may likewise be useful in practice, by assisting to make the quality of the venom well understood in these cases. Such an investigation, well handled, may likewise serve, as will be seen in the sequel, to explain the action of the venom of the viper on animals in general.

Superficial Wounds of the Skin.

I sat out by making the following experiments. I cut the hair with scissars from the skin of a part of the leg of a guineapig, and rubbed a portion of this, of about half an inch in length and breadth,

3   several

feveral times, with a fmall file. The fkin became red, and an almoft imperceptible quantity of blood exuded from it, which could not, however, form it-felf into entire drops. Having wiped it well, I poured on it with a large drop of venom, which, to make it flow eafier, and to extend itfelf over the whole furface of the rafped fkin, I had united with a drop of water.

The animal did not appear to fuffer in the leaft, and there was fcarcely any perceptible mark of ci-catrice. On the following day, obferving it to con-tinue found and vigorous, I had it bit twice in the foot by a viper. It died at the end of twenty-four minutes. This experiment I repeated twice, with nearly the fame refult; both guineapigs died after being bit.

I fhaved the hair with a razor, from the external lateral part of a guineapig's leg. The fkin was red, and a little moifture exuded from it, which was likewife of a reddifh tinge. I put two drops of ve-nom on this part, the fize of which was about two thirds of an inch. The animal did not fuffer the fmalleft inconvenience, and the fkin dried without efchar or cicatrice. On having it bit in the foot the next day, it died at the end of twenty-fix minutes.

I removed the hair with boiling water from a por-tion of the back of a guineapig, and made two very fmall, but very deep, incifions in it, wiping away the blood that flowed from them. I applied two drops of venom, unmixed with water, to the incifed fkin, which was eaten away for half its thicknefs,

by

by a wound that formed over the whole furface the venom had touched. This wound difcharged pus, and the next day was covered with an efchar. The animal was perfectly recovered in fix days, and on the feventh, on having it bit by a viper once in the foot, it died at the end of forty minutes.

I repeated this experiment, with the fame cir-cumftances, as nearly as I could judge, on two other guineapigs. The effects were exactly the fame; wound, fkin confumed for half its fubftance, pus, efchar, and recovery. On having them afterwards bit in the foot, they both died in lefs than an hour.

I likewife wifhed to make a fimilar experiment on an animal with a fkin much thicker than that of a gui-neapig. I chofe a very fmall rabbit, and removed the hair with a razor, in fuch a way, that there was a fenfible difcharge of blood. I applied to this part, about half an inch in length and breadth, two drops of venom. A true wound formed, and the fkin was entirely confumed, and covered with a great deal of pus. The rabbit notwithftanding did not feem to fuffer much, and at the end of feven] days was perfectly recovered. I had it twice bit in its leg by a viper, and it died at the end of fix hours. I repeated the fame experiment on two other rabbits, with the fame fuccefs.

The following conclufions may, I think, be drawn from the above experiments :

I. That the venom of the viper, applied to the fkin of guineapigs and rabbits, flightly fcraped or punctured, is not mortal.

II. That

II. That it produces but a flight difeafe in the fkin of guineapigs, and a fomewhat greater one in that of rabbits.

III. That this difeafe is confined to the part of the fkin touched by the venom.

I was defirous of making a fomewhat different experiment on the fkin of guineapigs, and accordingly removed the hair with fciffars from a portion of the back of one of thefe animals, of about the breadth of half an inch. I then made an incifion with a lancet, fo as not to puncture it through, but only for about half its thicknefs, and applied two drops of venom. A wound, occupying the whole fpace covered by the venom, formed, and fuppurated very abundantly, and the fkin was entirely confumed, and covered with a fcar. The animal did not appear to fuffer otherwife, ate conftantly, and recovered at the end of ten days.

This laft experiment feems to indicate that when the wounds of the fkin are deep, the effects of the venom, or its difeafe, are more confiderable, although not mortal ; and likewife, that the difeafe is entirely confined to the fkin,

Wounds in the Skin, through its whole fubftance.

I pinched the fkin of a fmall rabbit's leg with my thumb and finger, and pierced it five or fix times with a viper's tooth, from which the venom flowed. At the end of twelve hours, an encyfted tumour, filled with matter, formed in the fkin, an inch below the

wounds.

wounds. [The cyft was quite excoriated and bare of hair, and a little moifture exuded from it. The rabbit died on the fifth day.

I repeated this experiment on a rabbit of the fame fize, pricking the fkin feveral times with a venomous tooth. At the end of ten hours, the fame kind of tumour formed in the fame place; on the fecond day the fkin fell off; on the third the tumour burft; and the rabbit died four hours after.

I treated two other fmall rabbits in the fame way, and the effect was perfectly the fame. In both of them a tumour formed, and burft; and both died.

I had the fkin of a guineapig's back bit repeatedly by a viper, raifing it with pincers, to prevent the mufcles beneath from being wounded. In lefs than two hours, the part that had been bit became livid, and the animal died at the end of thirty-two hours, without an open wound. The fkin was gangrened, and the blood black and extravafated in the adipofe membrane, as far as the mufcles of the abdomen and breaft.

I repeated this experiment with the fame circumftances on four other guineapigs, all of which died. Neither of them had any wound, but the adipofe membrane had a gangrenous appearance, and was filled with black extravafated blood. The extravafation was extended to the adipofe membrane which covers the pectoral and abdominal mufcles, and was in fuch a quantity as to form a bag.

Experiments

Experiments on the Adipose Membrane.

The preceding experiments not only relate to the skin, but likewise to the adipose membrane. Whenever the tooth pierces through the whole substance of the skin, the venom must necessarily communicate itself to this membrane; and its effects, or the disease it occasions, will be communicated to both parts. It was therefore necessary to have the adipose membrane wounded apart, to know what related to the skin in the above experiments. It is not very easy to do this with precision and nicety.

I made an incision in the skin of a guineapig, near the groin, and introduced a drop of venom without its touching the skin. It brought on a tumour of the groin, which increased for two days. The third day the animal died. On opening the tumour, I found it filled with a great quantity of black, dissolved, and extravasated, blood.

I repeated this experiment on two more guineapigs, one of which died, the other did not. This last had scarcely any tumour. The one that died had a large one; with the same symptoms as in the preceding experiment. Two days after, I opened the one which survived, and which appeared sound, and in good health. I found the adipose membrane somewhat bloody, and with some humours extravasated in it; but all this in a slight degree. There was no appearance that could induce one to conclude, that the animal would afterwards have died

of

of the difeafe of the venom. It was vigorous, fed
well, and ran about in good health, whilft the other
was in the heighth of its difeafe at the end of two
hours after being bit.

These experiments ftill leave us in a doubt whether the venom might not have been communicated
to the incifed edges of the fkin. To obviate this, I
fell upon feveral modes of experimenting, but invariably met with difficulty in the attempts, and
fomething equivocal in the confequences.

After feveral trials, I purfued the following method :—I cut away a large portion of fkin from the
back of a guineapig, dried the adipofe membrane
well, and applied to it two drops of venom. The
circular piece of fkin I removed was more than an
inch in diameter. I fpread the venom on the membrane for about three lines in circumference, and at
equal diftances at all fides from the fkin.

In lefs than fix hours, the adipofe membrane became black as ink, and at the end of twelve it was
covered with an efchar, which continued fo long as
twenty-two ; the animal recovered notwithftanding.

I repeated this experiment on fix fmall rabbits,
and fix fmall guineapigs, and the confequences
were fomewhat different from each other.

In the firft place it muft be remarked, that neither of thefe animals died. Six of them were very
much difeafed, and recovered very late. Four had
flight fymptoms of the difeafe of the venom, and
feemed to be perfectly recovered at the end of the
fecond day. The others had no certain fymptoms

of

of difeafe. I think it may be faid, in a general way, that the venom of the viper is not mortal, if it penetrates no farther than the adipofe membrane.

Experiments on the Mufcles.

I ftripped the exteriour mufcles of a pigeon's leg of the fkin and adipofe membrane, without producing any fenfible hemorrhage. I introduced into one of thefe mufcles a viper's tooth filled with venom. A minute after the pigeon fell forward, and died at the end of ten. The wounded mufcle was extremely livid, throughout almoft the whole of its fubftance.

I repeated this experiment on four other pigeons, all of which in lefs than two minutes fell forward. One died at the end of eleven minutes; another at the end of feventeen; the third in a quarter of an hour; and the fourth not till four hours.

I ftripped feveral mufcles of the leg of a middle-fized rabbit of the fkin and adipofe membrane, and wounded them feveral times with venomous teeth, (a) in fuch a way that they entirely entered the muf-cles. I wounded them at the parts where there did not appear any confiderable veffels. There was fcarcely any difcharge of blood from the mufcles, which, notwithftanding, very foon became livid at

(a) Thefe are viper's teeth detached from the animal, but ftill adhering to the veficle filled with venom. I have already explained the method I purfue in experiments of this kind.

the

the places I had wounded. The animal not only furvived, but difcovered no figns of fuffering any great inconvenience; and at the end of fifteen hours there was fcarcely any difcoloration in the wounded mufcles. At the end of thirty hours, nothing was to be feen but the mechanical wound of the fkin, where the incifion had been made to come at each of the mufcles.

I repeated this experiment, with the fame circum-ftances, on another rabbit. The mufcle became difcoloured, but not much; and the animal, at the end of twenty-three hours, feemed to be free from all complaints, except that there ftill remained a folution of continuity in the fkin.

I entirely ftripped feveral mufcles of a guineapig's leg of the fkin and adipofe membrane, and plunged a tooth, charged with venom, betwixt the fibres in fuch a way, that few or no veffels were divided. The mufcle became livid, but the animal reco-vered.

I repeated this experiment on the bared mufcles of feveral other animals, fuch as guineapigs and rabbits, and found that in thefe cafes the venom of the viper does not fail to bring on a complaint, which, although it is frequently very violent, is never mortal.

The Venom of the Viper, when simply applied to the Muscular Fibres, is entirely innocent.

I wished to know what would be the effects of the venom, when simply applied to the muscles, without cutting the fibres.

I stripped the muscles of a pigeon's leg of the skin, and contrived in such a way, that the uncovered fibres and vessels did not bleed sensibly. The experiment succeeded so well, that the muscles, stripped of the adipose membrane, appeared perfectly dry. On these muscles I laid a large drop of venom, observing that it did not communicate itself to the adjacent parts. The pigeon had no complaint, and the wound I had made healed very soon.

I got ready another pigeon in the same way, but took care that the muscles should bleed a little; one vein in particular bled considerably. I applied the venom, and the pigeon died at the end of thirty hours, with a very slight change in the parts that had been wounded.

I repeated this experiment on four other pigeons, the muscles of which did not bleed. Neither of them died, nor seemed to have any other complaint than that occasioned by the incision in the skin.

When we know how small a quantity of venom is capable of killing a pigeon, as it were, instantly, we cannot hesitate to pronounce, that the venom of

the

the viper, when fimply applied to the mufcular fibres, is entirely innocent.

The Venom of the Viper does not lofe its deadly Qua-lities, even after it has acted on an Animal as a Poifon.

I was defirous of feeing whether the venom of the viper, after having communicated the difeafe to one animal, would act as a poifon on another. To affure myfelf of this, I laid the mufcles of a pigeon's leg bare, and made fmall incifions in them, into which I introduced about a drop of venom.

I likewife got ready another pigeon, making fmall incifions in its mufcles, as I had done in thofe of the firft. At the end of four minutes I put the bared mufcles of the two pigeons in contact, and kept them in that ftate for two minutes. Neither of the pigeons died: the firft, however, was very ill; the fecond had fcarcely any complaint.

I laid the mufcles of two other pigeons bare, and made fmall incifions in them. I wounded thofe of one with a venomous tooth, and at the end of four minutes put them in contact with thofe of the other, keeping them together in this way for three minutes. The firft pigeon died at the end of three minutes more; the fecond at the end of an hour.

I repeated this laft experiment on two other pigeons. The one that was venomed by the tooth
died

died at the end of eight minutes; the other at the end of eighteen.

Consequently the venom of the viper, as in all the cases related above, continues to be such, and does not lose its deadly qualities, when it mixes with the blood of living animals, and excites in them the usual disease.

Animals bit in the Breast.

I had a pigeon bit once by a viper in the breast. I treated it, and it died at the end of ten minutes.

I had another pigeon bit twice in the breast by a viper, and treated it. It died at the end of two hours.

I had six pigeons bit in the breast by as many vipers, each twice by a distinct viper. Three were treated, and three not. They all died; the three that were treated at the end of 10, 20, 50, minutes; the other three at the end of 17, minutes, and 2, 4, hours.

I had six others bit an equal number of times; three in the breast, and three in the leg. They all died; the three bit in the leg at the end of 10, 15, 20, minutes; the three in the breast at the end of 17, 50, minutes, and two hours.

These few experiments on pigeons would lead one to suspect, that bites in the breast are not more dangerous than those of the leg; and that it may even be the reverse. They are however not sufficient

cient in number to admit of any certain confequences being drawn from them.

I had a guineapig bit twice in the breaft by a viper, and immediately treated it. It died at the end of two hours.

I had another guineapig, of a much larger fize, bit twice in the breaft by a viper, and treated. At the part where it was bit, a large wound, which continued open for upwards of fifteen days, form-ed; the guineapig at length recovered.

I had a very large guineapig bit twice in the breaft by a viper, and treated it immediately. It had no fymptom of difeafe. Two days after I had it bit afrefh by another viper, in the fame place, and at the end of twelve hours it died.

The fkin of guineapigs, particularly that part of it which covers the breaft, is very tight, in confe-quence of which the viper finds it very difficult to feize it betwixt its teeth. I was feveral times de-ceived by this, fuppofing the animal bit when it was not; and was therefore obliged to repeat the experiment.

I had a fmall rabbit bit in the breaft by a viper, and immediately treated it. At the end of thirty feconds it fell on its belly, and died in lefs than a minute.

I had another rabbit, of the fame fize, bit in the breaft, and did not treat it. It had a fmall wound, and recovered at the end of three days.

I had four rabbits bit in the breaft, each twice by a diftinct viper. Two were treated, and two not.

The

The two that were treated died; one at end of an hour, the other at the end of ten. Of thofe not treated, one died in an hour, the other had a very fmall wound in the part bit.

I had a fowl bit twice by a viper in its breaft, towards the right wing. I treated it, and it died at the end of twenty-four hours.

I had another fowl bit twice by a viper in the fame place, and did not treat it. It died at the end of nine hours.

I had four other fowls, like the preceding ones, bit, and obferved the fame circumftances. They all four died in eighteen hours.

I had four other fowls bit, two in the breaft, and two in the leg. The two that were bit in the breaft died in lefs than ten hours. One of thofe bit in the leg died at the end of twenty-feven hours; the other was violently difeafed, but recovered.

Had the number of experiments been greater, we might have deduced from them, that to fowls the bite of the viper in the breaft is more dangerous than that in the leg. This is contrary to what was obferved in the rabbits and guineapigs.

Animals bit in the Belly.

I had a rabbit bit twice in the belly by a viper. At the end of eighteen hours a very large tumour formed in the part bit. This tumour encreafed for four days, and the hair fell from the fkin, which

was

was corroded and ulcered. The animal, nowith-
ftanding, lived twenty days.

I had another rabbit, of the fame fize, bit repeatedly in the belly by a viper. At the end of twelve
hours a tumour formed, and the hair and epidermis
came away. The tumour was moift and bloody,
and burft at the end of eighteen hours, when an
ulcer formed, of two inches and an half in length,
and more than an inch in breadth. The rabbit furvived, but it was more than twenty days before it
recovered.

I had two others bit in the belly in the fame way.
Both of them had a tumour, which was fucceeded
by an ulcer that remained open for feveral days;
and both recovered.

I had two other rabbits of the fame fize bit feveral times in the belly by two vipers. One died at
the end of twenty-fix hours; the other had a wound
which covered the whole of the fkin of the lower
part of the belly, and continued ill twenty-fix days.

Experiments on the Inteſtines.

I opened the belly of a rabbit, and had the *ileum*,
at the diftance of three inches from the *colon*, bit
twice by a viper, binding up the part as well as I
could. · The rabbit died at the end of fix hours.
The inteftine was inflamed, black, and contracted,
more than fix inches above and below the part that
was bit; fo that thefe changes had extended to the
colon.

colon. The *mefenterick veffels* were black and fwelled, and the blood curdled.

I repeated this experiment on four other rabbits, each of which I had bit in the inteftines in the fame way by a viper. The refult of thefe experiments was perfectly analogous to that of the former one.

Experiments on the Liver.

Having opened the belly of a rabbit, I wounded the right lobule of the liver, in the inner part, with a venomous tooth. At the end of a few feconds, the creature began to cry and to writhe itfelf, and died in lefs than two minutes. All the veffels of the liver were filled with black and clotted blood; the mefentery was in the fame ftate; and the heart and auricles were filled with black, but fluid, blood.

I wounded the outer lobule of the liver of another rabbit in two places with a venomous tooth. The creature drew itfelf together, but did not cry. It died an hour after.

I introduced a venomous tooth into the outer lobule of the liver of a third rabbit, and did not withdraw it immediately. This one cried, writhed itfelf, and died in lefs than a minute and an half. The blood was coagulated both in the liver and mefentery.

I introduced a venomous tooth in the ufual way into the inner lobule of the liver of two other rabbits, and kept it there for fome time. Thefe ani-

O   mals,

mals, as ufual, cried out after a few feconds, and died in lefs than two minutes. The blood in the liver was black and coagulated; that in the heart and auricles was likewife black, but in a fluid ftate. I did the fame to the outer lobule of the liver of two rabbits, but withdrew the venomous tooth immediately after having introduced it. One began to cry and writhe itfelf after a few feconds, and died in two minutes. The other lived nearly two hours. The blood in the liver of the firft was quite coagulated; as it was in a degree in the fecond. In the former, the blood in the auricles and ventricles was fluid; in the latter it was coagulated.

Experiments on the Ears.

I had the ear of a middle-fized rabbit, towards its extremity, or point, bit twice by a viper. The part was a little fwelled at the end of fix hours; the rabbit, however, ate, and was lively. At the end of four days it was perfectly recovered.

I had two other middle-fized rabbits bit in the fame way at the extremity of the ear, each twice by a diftinct viper. The ears fwelled a good deal, but the rabbits ate, and were lively. At the end of five days they were both recovered.

I had another rabbit bit in the right ear, towards its extremity, twice by a viper. I treated the part, in which there was a confiderable fwelling that did not fubfide till after fixteen days.

I had

I had a rabbit's ear bit twice by a viper, at a third of its length from the bafis. At every hole made by the teeth in the oppofite fides of the ear, a drop of blood appeared, and befide it a fmall drop of venom, which, although it was in contact with the blood, did not unite with it in the leaft. There were four holes made by the teeth at each fide the ear, fo that the fmall drops of venom were eight in number. The ear fwelled a good deal, and the fwelling did not fubfide till after twenty days.

There is no difficulty in accounting for the fmall drops of venom that appeared at the oppofite fides of the ear. We know that the venom flows from the point of the tooth. The ear of a middle-fized rabbit is not fo thick as the viper's tooth is long, which muft of courfe pierce the ear through. When the viper withdraws the tooth, the venom has already reached the point of it; and from the elafticity of the fkin of the ear, which clofes the hole it went out at, is forced to fhed itfelf at the fides of it. In finding its way to the part of the ear at which it entered, the tooth in the fame way leaves the venom, which it continues to fhed, at the edges of the hole at this fide. I have fince obferved thefe fmall drops of venom on each fide the ear in almoft all the rabbits I have had bit in this part, and find them in general to be larger at the part the tooth went out at, than at that where it entered; particularly if the viper is prevented from withdrawing its teeth too fuddenly.

O 2

I had

I had a rabbit bit in both ears at a third of their length from the bafis. Each ear was bit three times by a diftinct viper, and both of them fwelled violently, for nearly eight lines towards the bafis. The rabbit was very much difordered, and did not eat for feveral days, when it began to feed fparingly. It was not perfectly recovered till the end of twenty days, and was then very much wafted.

I had two others bit repeatedly at the fame part of the ear, by two vipers. At the end of the fecond day, the ears were disfigured by a fwelling, which became fo large, that in two days more they hung down on each fide the neck. One of the rabbits died at the end of eight days, with its ears ulcered and fphacelated; the other recovered, but not till the end of twenty-eight days.

I had a middle-fized rabbit bit once in the ear by a viper. The ear bled a little, and two fmall drops of venom appeared at the fides of the two holes made by the teeth. I did not treat it. There was a degree of tumour and inflammation in the part, and at the end of thirty hours the rabbit was perfectly recovered.

I had another rabbit bit, of the fame fize as the preceding one. I treated it immediately, and made it fwallow the volatile alkali. The ear fwelled exceedingly, and became livid at the part where it was moft fwelled. The tumour continued fix days, and in four more the animal recovered.

I had four rabbits bit in the ears by as many vipers. Two were treated, and two not. Neither of
them

them died. The ears in all of them fwelled confiderably, and they all recovered at the end of three
days.

Having affured myfelf that the bite of the viper
in the ears of rabbits is not very dangerous, I thought
of having thefe animals bit by feveral vipers, in different parts of the two ears. For this purpofe I
chofe a dozen middle-fized rabbits, and had them
all bit repeatedly in feveral parts of each ear, and
each by three vipers. The parts fwelled exceedingly, and continued in that ftate for upwards of
twelve days. Three of the rabbits had an enormous bag or tumour in the fore part of the neck,
larger than the head itfelf. Thefe tumours were
filled with a humour, and yielded to preffure. At
the end of two days they increafed in fize, and the
ears became ulcerous. The rabbits recovered in
fixteen days.

Experiments on the Pericranium.

I laid bare the pericranium of a pigeon, by removing a good portion of the fkin, and made feveral fmall incifions into it with the point of a lancet.
I poured venom upon it, but in fuch a way that it
did not reach to the adjacent parts of the integuments that had been cut. The pigeon did not feem
to be at all difordered by it, and recovered in the
fame fpace of time as another did, which I had prepared by way of comparifon, and to the pericranium
of which I had not applied the venom.

I re-

I repeated this experiment on four other pigeons, with the fame fuccefs. Neither of them died, and in neither was the attack of the difeafe of the venom perceptible.

On the Bones and Periofiæum.

I laid bare the cranium of a pigeon, ftripping off a good part of the pericranium. I made fmall wounds in the cranium with a lancet, taking care not to pierce the whole fubftance of it, and introduced a confiderable quantity of venom, preventing it as ufual from communicating to the neighbouring parts. The animal not only furvived, but did not appear to have fuffered the fmalleft inconvenience.

The confequences of three experiments on pigeons, treated in the fame way, were the fame.

Having laid bare the *tibia* of two pigeons, and freed it well of the cellular membrane, I wounded both periofiæum and bone in feveral places with the point of a needle, and poured the venom copioufly upon them. The pigeons had not the fmalleft perceptible complaint, and recovered in the fame time as two others did, that I had treated in the fame way, but without applying the venom, to ferve as a comparifon.

I repeated this experiment with the fame circumftances on two other pigeons, and the refult was exactly the fame. Neither of them died, nor had the fmalleft fymptom of the difeafe of the venom.

I laid

I laid bare the perioftæum of the *tibia* of fix other pigeons, and having pierced it in feveral places with a needle, moiftened it with venom. Neither of the pigeons died, nor had any complaint.

Dura Mater and Brain.

I removed a portion of the cranium of a pigeon, taking care to lacerate the dura mater as little as poffible. I wiped this membrane, which was well expofed, as gently as I could, with dry lint, and applied to it a drop of venom. The pigeon had no fymptom of the difeafe of the venom, and recovered within the fame fpace of time as another did, that I had prepared in the fame way, but without applying the venom, as a comparifon.

This experiment, on two other pigeons treated as above, terminated in the fame way.

I removed a portion of the cranium of a pigeon, and made incifions in the dura mater all round, introducing at one of the apertures a drop of venom The pigeon recovered, without having had any fymptom of the difeafe of the venom.

After having removed the dura mater of another pigeon, I made a flight incifion into the brain, and introduced the venom. The animal recovered in the fame way with the preceding one.

A third pigeon, on which I made the fame trial, died at the end of four hours.

Marrow

Marrow of the Bones.

I cut the *tibia* in two pigeons, towards the lower
extremity, and introduced lengthways into the marrow two fmall bits of wood covered with venom.
Neither of the pigeons died, nor had any fymptom
of difeafe.

I cut the *tibia* in the fame part, of two other pigeons, and introduced into the marrow two fmall
bits of wood, well covered with venom, keeping
them there fix minutes. Neither of the pigeons
had any apparent fymptom of the difeafe of the venom.

I repeated this experiment with the fame circumftances on four other pigeons. Each of the trials
ended the fame way, the pigeons all recovering
within the fame fpace of time that two others did,
which I employed as a comparifon, without venom-
ing them.

The Venom applied to the Tranfparent Cornea.

I pierced the tranfparent cornea of the right eye
of a large rabbit with a venomous tooth. The
aqueous humour flowed out. I then, with another
venomous tooth, firft fcratched, and afterwards
pierced, the tranfparent cornea of the other eye. At
the end of an hour I found the right eye filled with
the aqueous humour, and perfectly found. At the
end

end of eighteen hours, a fmall white fpot formed in the tranfparent cornea of the other eye, but without any inflammation about it. At the end of three days, the white fpot in the left eye was raifed above the cornea.

I fcratched the cornea of another rabbit with a tooth well dried, and at length pierced it. At the end of fourteen hours a dark fpot appeared, and two days after the cornea was raifed up in the form of a pearl.

I poured a drop of venom into the eye of a large rabbit, which I examined every hour. At the end of eighteen hours, the *membrana nictitans* feemed fomewhat redder than ufual.

I poured two drops of venom into the eye of another rabbit, and this was not fucceeded by any inflammation.

I made the fame experiment on the eye of a third, which continued in its natural ftate.

I repeated it on three other rabbits, neither of the eyes of which became fenfibly inflamed.

I moiftened the eyes of a large rabbit feveral times with a confiderable quantity of venom, and likewife applied feveral drops to its lips and tongue. At the end of three hours the *membrana nictitans* appeared a little red, but at the end of eighteen hours returned to its natural ftate.

I put feveral drops of venom on the tongue of another rabbit, and fmeared it on the lips and palate with a brufh. There was no fwelling in any

part

part of its mouth, neither was the rabbit at all difordered.

This experiment repeated on two other rabbits
was attended with the fame confequences. No part
of the mouth was either fwelled or inflamed.

C H A P T E R   V.

*Experiments on the Comb, Gills, Nofe, and Neck, of
Animals.*

My next purfuit was that of examining the effects of the venom of the viper on the comb, gills,
nofe, and neck of animals. My experiments on
thefe parts have been attended with unexpected and
interefting confequences; and therefore I have
thought it proper to treat them apart, in an extenfive way.

Experiments on the Comb of Fowls.

I had the comb of a fowl bit twice by a viper.
There was a confiderable hemorrhage from the
wounds made by the teeth. At the end of three
hours the gills were fwelled, and in fix a large tumour or bladder was formed. The fowl died at the
end of four days, without having either eat or drunk.

The

The tumour of the gills, which united them into one monftrous body, was filled with a wetry flefh-coloured humour, and with an heap or web of filaments and veffels.

I had a fmall cock bit once in the comb by a viper, and treated it immediately. It died at the end of ten minutes.

I had another cock, of the fame fize, bit once in the comb by a viper, and treated it. At the end of two hours both gills had already fwelled; at the end of twenty-two this fwelling was very much abated; and in thirty-fix there were only fome little remains of fwelling in one of them. In forty hours the cock was perfectly recovered.

I had the comb of a large cock bit three times by a viper. It was branched, pointed, and more than a third of an inch in thicknefs. It bled a little, and there were fome fmall drops of venom befide the holes made by the teeth. I made a fmall wound in the comb with the point of a lancet, and introduced a fmall quantity of venom. The cock had no fymptom of complaint. Two days after I had it bit twice in the comb by another viper. At the end of two hours the part appeared fomewhat livid towards its bafis, and perhaps a little fwelled. At the end of three hours the gills were very much enlarged, and at the expiration of twenty, were become of a mônftrous fize, and livid for their whole extent. At the end of twenty-three hours they burft, and the cock died very foon after.

There

There cannot be the fmalleft doubt but that the
venom in the firft cafe was thrown out by the blood,
and this happens not unfrequently. It is much more
difficult to account for the tumour that, notwithftanding the cock was bit in the comb, formed in
the gills. However I have frequently feen fomething fimilar happen in other animals. The bite
made in the leg of a rabbit frequently caufes
a tumour, or an obftruction of the humours in the
moft inferior parts of the fame leg. But the experiments muft be continued.

I had the comb of a fowl bit by two vipers, by
each twice. At the end of two hours one of the
gills only began to fwell. In twenty they were
both very much fwelled, and united in fuch a way
that they formed a fingle body. At the end of
thirty-fix they were enormoufly fwelled and very
livid. In ten days the fowl recovered. On the
fourth day of the difeafe it refpired with difficulty,
and with a hiffing noife; the glottis was inflamed
and open, and the trachea arteria fwelled.

I had the comb of another fowl bit feveral times
by two vipers. At the end of three minutes the
part next the head was livid, and appeared a little
fwelled. In an hour the livid colour and tumefac-
tion feemed to be fubfided, but on the other hand,
the gills were enlarged. In three hours one of the
lower eye-lids exuded blood from all its fmall orifices. The gullet and palate were black. In twelve
hours the fowl was in a dying ftate, the gills being

livid

livid and of an enormous fize.   It died at the end
of thirty-three hours.

I had the comb of a fowl bit feveral times by a
viper.   One of its gills fwelled a little.   At the end
of thirty-fix hours this fmall degree of fwelling had
difappeared, but the fowl refpired with difficulty,
and in doing this made a great noife.   The wind-
pipe was fwelled, and very much enflamed, even at
the end of fix days.   The animal was perfectly recovered in ten.

All thefe experiments fhow that there is an immediate communication of veffels and humours,
betwixt the comb and gills of fowls.   I do not
give a detail of more than ten experiments befides
that I made on fowls, fince they terminated in the
way with the cafes juft related.

Experiments on the Gills of Fowls.

I was defirous of knowing what would be the
confequence of having fowls bit by vipers, not in
the comb, but in the gills only; that is to fay, whether the bite would be equally dangerous, and whether the tumour would fly up to the comb without
forming in the gills, or would form both in the gills
and comb.

I had the gills of a fowl bit repeatedly by two
vipers.   At the end of two ininutes they had already fwelled, and become livid.   There was a
great flux of humours in the eyes, which were
clofed

clofed by the enlargement of the membrana nictitans.   In lefs than an hour the gills were of an
enormous fize, and livid all over.   The fowl died
at the end of the fifth hour.

I had the gills of a fecond fowl bit twice by a
viper.   They fwelled in lefs than four minutes,
and in two hours were extremely large and livid.
The comb appeared a little dark towards its points
and edges.   The fowl died at the end of three
days.

Thefe trials may induce one to fufpect that
wounds made in the comb are lefs dangerous than
thofe made with the fame circumftances in the
gills.

To come at the truth of this hypothefis, I made
the following experiments.   I had fix fowls bit,
each twice by a diftinct viper ; three in the comb,
and three in the gills.   One only of the former
died, and two of the latter.

On repeating this experiment on fix other fowls,
the refult was fomewhat different.   Only one of
thofe bit in the comb died, and all thofe that were
bit in the gills.

Thefe new experiments led me to think that my
conjecture was very probable ; that is to fay, that
the bite of the viper in fowls is more dangerous
when made in the gills, than when it was made in
the comb.

The accident which fupervenes in the fowls the
comb of which has been bit by vipers, is very fingular.   The action of the venom, or its difeafe, is

conveyed

conveyed to a remote part that has not been bit ;
but when the gills are bit, the venom does not fly
up to the comb, nor does the difeafe communicate
itfelf to that part, and yet the ftructure both of the
gills and comb is the fame, and the veffels and
nerves are common to both.

This circumftance ftruck me fo foreibly, that
I thought it deferving an analyfis of fome kind,
and of being fearched into by ftill further trials.

I began by having the comb of a fowl bit
once, and at the end of fifteen feconds cut off
both gills.   The fowl not only recovered, but
there was no change in the gills, nor any appearance nor fymptom whatever of the difeafe of the
venom.

I had another fowl bit once in the comb, and at
the end of fifteen feconds cut it entirely away. The
gills did not fwell, neither had the fowl any fymptom of the difeafe of the venom.

I had the gills of a large cock bit repeatedly by
a viper.   In fix hours they were both very much
enlarged.   On the following day they were ftill
more fo, and were befides livid.   The cock recovered at the end of thirteen days.

I had the gills of another cock, a very large one,
bit feveral times by two vipers.   At the end of ten
minutes I cut them off.   On the following day it
ate, and appeared in health, and after three days was
perfectly recovered.

I repeated this experiment on the gills of fix
other cocks, each of which I had bit repeatedly
2   by

by two diſtinct vipers. I cut off the gills in all of
them, but at different intervals; at the end of 1,
2, 4, 8, 16, 32, minutes. They all recovered, and
had no other complaint, than that produced by
the cutting off of the parts.

I had a large cock bit ſeveral times in the comb
by two vipers, and after eight minutes cut off its
gills. It died at the end of three hours.

I had another cock, a very large one, bit repeatedly in the comb by two vipers, and after four minutes cut off its gills. It died at the end of twenty-
ſeven minutes. It was ſcarcely bit by the firſt viper,
when it could no longer ſupport itſelf, or hold its
head erect. It opened its beak, from which a glutinous humour flowed, and breathed ſhort, and with
difficulty.

I repeated this experiment on ſix other cocks,
each of which I had repeatedly bit in the comb by
two diſtinct vipers. I cut off the gills in each of
them at the end of four minutes. Three died in
leſs than twenty hours; the other three were very
much diſeaſed, and did not recover till the end of
ten days.

Experiments on the Neck of Animals.

I had a ſmall guineapig bit twice by a viper in
the back part of its neck. I treated it. It died at
the end of forty minutes.

I had

I had a middle-fized rabbit bit twice by a viper in the upper part of the neck, and treated it. It died at the end of twenty-four hours.

I had two guineapigs bit in the neck, each twice by a diftinct viper. One was treated, and the other not. Both of them died; the one treated at the end of an hour, the other at the end of four.

I had two fmall rabbits bit in the neck, each repeatedly by a diftinct viper. I treated one, and made it fwallow the volatile alkali feveral times, and did nothing to the other. Both of them died; the firft at the end of four hours, and the other at the end of twenty-two.

I had a large guineapig bit twice in the neck by a viper. In an hour the part of the neck that had been bit became fwelled and livid. At the end of twenty-three hours a large wound appeared. At the end of the fecond day the humours which formed the tumour, had extended to beneath the chin, and formed a large bag or bladder. In four days the tumour had fwelled to fuch a degree, that it almoft covered the breaft. The fkin had loft its hair and epidermis, and a flightly-coloured humour exuded from it. At the end of fix days the fwelling began to diminifh, and the guineapig recovered at the end of fifteen.

The difeafe in this animal, or the matter which defcended from the upper to the lower part of its neck, and which even reached to the breaft where it formed a cyft or bladder, bears a ftrong analogy to the circumftances that were obferved on having

VoL. I.　　　　　P　　　　　the

the fowls bit. There is only this difference, that
in fowls this effect is more frequent, and is oftener the cafe than otherwife ; whilft, on the contrary, it happens very rarely in quadrupeds bit in
the neck, at leaft in guineapigs. Of twenty-two
animals bit in the fame way, of which eleven were
treated and eleven not, I found five in which this
tumour defcended below the neck, and formed a
bladder. Of thefe five, three were treated, and
two not. The number of deaths, which confifted
of four in the whole, was equal on both fides.

It is however certain, that having had fome others
bit, but each of them by feveral vipers, and feveral
times by each, the tumour or cyft formed in the
interiour part, in a greater number of them, and that
almoft all of them died.

The confequences were analogous, on trying the
fame experiments on rabbits. The cyft fometimes
forms beneath the chin of thefe animals, although
they have only been bit in the neck ; and this happens more frequently when they have been bit by
feveral vipers, in which cafe they die much
readier.

Experiments on the Nofe of Animals.

It remained for me to examine the bite of the
viper on a part that is held the moft fenfible, and
the moft likely to occafion death when it receives
an injury, in fome particular animals.—This is the
                                                    nofe.

nofe. It appears, that the cat, an animal very obftinate in dying, perifhes as readily as others, if
ftruck in this part.

Mead reckoned it fo fenfible and fo dangerous
in dogs, that, wifhing to affure himfelf of the efficacy of a remedy againft the bite of the viper, he
had a dog bit on the nofe, and applied the remedy.
The dog lived, and this was fufficient to give the
remedy the reputation of a true fpecifick; fo ftrong
was the opinion, that a bite of the viper on the
nofe was mortal.

I fhall not relate here all the experiments I made
on this part, but only a fmall number of them,
which will be fufficient to give a clear idea of the
fallacy of fome popular opinions.

We fhall fee what we ought to think of the bite
of the viper on the nofe, and how abfolutely neceffary it is to confult nature by facts and experiments.
Nothing is more dangerous and more uncertain in
refearches of this nature, than a vague analogy, or
a feducing and probable reafoning. Nature is not
to be divined, and *prophets* in the fcience of phyficks
are not to be believed.

I had a fmall rabbit bit twice on the nofe by a
viper. In two minutes the part was fenfibly inflamed. In three hours a tumour was formed in
the neck, beneath the chin. In feven hours this
tumour was become very large.—The animal recovered however.

I had another rabbit, fomewhat larger than the
former one, bit on the nofe by a viper, and treated

it.

it. It was bit twice, but one of the bites was made on the upper lip, at the fide of the nofe. In the fpace of two minutes the nofe was fwelled, and a very large tumour formed under the chin. At the end of twenty hours this tumour burft, and dif-charged a great deal of matter. The rabbit reco-vered at the end of fix days.

I had a third rabbit, of a middle fize, bit twice on the nofe by a viper. In a very little time the part fwelled and inflamed. In two hours a tumour formed beneath the chin, which at the end of feven difcharged blood, and was very large. In thirty-fix hours the tumour and fkin began to dry, and the animal recovered at the end of the fixth day.

Six other rabbits were bit in the fame way. Nei-ther of them died, and the effects of the bites were pretty much the fame as thofe related above. The bite of the viper on the nofe of rabbits, contrary to what one would naturally have thought, feems to be lefs dangerous than that in other parts. The difeafe it produces, as to the feat of it, is very fimi-lar to that in the comb of fowls. Here again a tu-mour forms, in a part where the animal has not been bit, and beneath the feat of the bite, in which, in moft cafes, the venom fcarcely occafions a real and fenfible complaint. The only effential dif-ference is, that the tumour in rabbits is of a greater extent, reaching fometimes to the middle of the breaft.

We are now to fee, whether the fame thing hap-pens in animals of other fpecies'.

I had

I had a large guineapig bit on the nofe by a viper. In two hours the part was very much fwelled; at the end of three the fwelling was diminifhed, but in its place a large tumour formed beneath the chin. At the end of fifteen hours, the tumour broke, and difcharged a great deal of blood and ferum. In thirty-fix hours the difcharge ceafed, and the opening in the fkin dried up. The animal was perfectly recovered at the end of four days. It was never very much difordered, fince it ate during the whole time of its illnefs.

I had another large guineapig bit twice on the nofe by a viper. The nofe and mouth fwelled very much, but this fwelling diminifhed in proportion as a tumour formed under the chin. After twenty-two hours the tumour, which had broke an hour before, began to dry up, and at the end of thirty-fix feemed perfectly dry. At the end of two days the animal was recovered. During the whole courfe of its complaint it fuffered but little, and ate conftantly.

I had a large guineapig bit on the mouth by two vipers, each biting twice. The nofe fwelled in lefs than three minutes, and ftill more fo at the end of ten. Two hours after, a tumour formed beneath the chin, when the fwelling of the nofe diminifhed, and in a fhort time entirely fubfided. At the end of twenty-three hours the tumour beneath the chin was fo large as almoft to cover the breaft, and in two hours more it burft. In the fpace of five other hours the animal recovered.

I re-

I repeated this experiment on another large guineapig, which I had bit by three vipers, each viper
biting three times. The nofe and mouth both
fwelled very much, but continued in that ftate only
four hours. At the end of the fecond hour a large
tumour appeared under the chin, which in twenty-
three hours was become enormous, and reached to
the breaft. This tumour broke at the end of thirty
hours, but the animal was not perfectly recovered
till the eighth day. The bones of the nofe were
laid bare, and the furrounding fkin all confumed.

I made the fame experiment on two other guineapigs, but fmall ones. One died at the end of twelve
hours; the other had the ufual tumour, was exceedingly ill, but did not die.

The bite of the viper on the nofe feems to produce pretty much the fame effects on guineapigs as
on rabbits; and it appears that the venom is lefs
dangerous in this part than in any other. The
fame effects are here conftantly obferved as to the
feat of the difeafe; but are they the fame in all
other animals?—I fhall relate what I met with in
dogs and cats, creatures that enter into the plan of
my prefent refearches, and this will fhow how little
analogy alone ought to be trufted, and that the
fame caufe produces very different effects, on
fimply changing fome circumftance, which one
would not have fuppofed capable of influencing a
great deal.

I had a fmall dog bit repeatedly on the nofe by
two vipers. Both nofe and mouth fwelled, and the
                                                   dog

dog died at the end of eight hours, without any symptom of difeafe in any other part.

I had another dog, twice as large as the pre-ceding one, bit repeatedly on the nofe by two vipers. Its mouth fwelled to fuch a degree, that the lips were very much enlarged twelve hours after. It vomited feveral times, and continued ill for three days, when it began to drink. On the fourth it ate, and on the fifth was perfectly reco-vered.

I had another dog, ftill larger than the one laft mentioned, bit on the nofe by three vipers, each viper making three bites. In a little time its nofe, mouth, and lips, were fwelled fo as to become hi-deous. It vomited a great many times, ate and drank on the fourth day, and recovered on the fifth.

I had another dog, of the fame fize as the pre-ceding one, bit on the nofe by four vipers, each biting three or four times. It had a bite at the fide of its nofe, and another on one of its lips. It vo-mited frequently, neither ate nor drink till after the third day, and recovered on the fifth.

I had another large dog bit on the nofe by fix vipers, each of which bit three or four times. Its nofe and mouth fwelled enormoufly; it vomited a great many times, ate after the fourth day, and recovered on the fixth.

Laftly, I had another dog, of the fame fize as the three preceding ones, bit on the nofe by fix vipers, by each three or four times. The part fwelled vio-

lently,

lently, and the animal did not eat till after the fifth day. It vomited frequently, and recovered at the end of seven days.

Rabbits and guineapigs bit on the nose, usually have the disease beneath the chin, and not in the part bit. It is quite the contrary with dogs, in which the disease is entirely confined to the nose, the part that received the bite. They therefore form an exception to the cases related prior to theirs.

It is likewise singular, that as the action of the venom is confined to the nose, it does not produce incurable wounds and gangrenes in that part. We, however, find it to be quite otherwise;—bites on the nose in dogs are very rarely attended with a wound in the part, and the animal not only makes a strong resistance to the disease, but the latter at the same time appears to be very slight, since the recovery takes place in a few days.

Experiments on Cats bit on the Nose.

We have seen above, that the cat makes the strongest resistance of any animal to the bite of the viper, although the venom constantly produces in it a disease. We may therefore conjecture, that this bite on the nose of cats will not be productive of mortal effects. But we know, on the other hand, that mechanical percussions on the nose are

dangerous

dangerous to thefe animals, and that they foon die, if they fall from a height on this part.

From thefe confiderations, I wifhed here again to have recourfe to experiments, which can alone determine the truth.

I had a middle-fized cat bit repeatedly on the nofe by a viper. Its mouth fwelled for a confiderable extent. It ate on the fecond day, and recovered on the third.

I had another, of the fame fize, bit repeatedly on the nofe by a viper. A few minutes after, the part fwelled. The cat vomited twice, ate on the fecond day, and was perfectly recovered on the third.

In this fecond cat, the difeafe of the venom was fo very flight, that the animal appeared to fuffer but little during its continuance.

I had a third cat bit repeatedly on the mouth by a viper. One of the bites fell on its upper lip, which bled a good deal, and the whole of its mouth fwelled very violently ; however it ate on the fecond day, and on the third was recovered.

I had a large cat bit repeatedly on the nofe by a viper. The part bled very much, and fwelled a few minutes after. At the end of twenty hours it was ftill more fwelled, notwithftanding which, the cat appeared but little difordered. It recovered at the end of forty hours.

I had another cat, of a middle fize, bit repeatedly on the nofe by a viper, which likewife bit it on the mouth and lips. The mouth fwelled at the end of five minutes, and at the end of five hours

the

the cat vomited several times.   In thirty-six hours
it was perfectly recovered.

I had another cat, of a middle size, bit on the
nose, and on the mouth both below and above.
After seven hours it vomited several times.   Its
nose and mouth were but little swelled, and at the
end of twenty hours it recovered.

Another cat, of a middle size, was bit by three
vipers, each of which bit three times, or upwards,
on the nose, mouth, and even within the palate,
from which there was an hemorrhage.   Some minutes after, its mouth swelled, it vomited several
times, but the palate did not swell at all.   It ate at
the end of three days, and at the end of the fifth
was perfectly recovered.

I had another cat, somewhat larger than the preceding one, bit by four vipers.   Each viper bit several times, on the nose, mouth, and lips, and in
the palate, insomuch that the cat, feeling one of the
bites within its mouth very sensibly, seized the
viper betwixt its teeth, and almost severed its head
from its body.   The nose and mouth in this cat
swelled very much, it vomited several times, ate
on the fourth day, and recovered on the sixth.

I repeated these experiments on three other cats,
which I had bit repeatedly in the nose by a viper,
and the effects were pretty much the same.   We
may therefore, I think, conclude, that the bite of
the viper on the nose is not very dangerous to dogs,
and that it is still less so to cats.

It

It is, however, very ftrange, that in both thefe animals there is no tumour beneath the chin, and that the local difeafe is confined to the part bit; whilft, on the contrary, the difeafe in rabbits and guineapigs is not in the part bit, but in another part of the animal beneath it.

It is clear, that this difference can only depend on the different organization and nature of thefe animals ; and it is precifely this diverfity that we are ignorant of.

I muft here obviate a difficulty that may be made by thofe who are not accuftomed to fuch experiments.

Thefe may oppofe, that bites in the nofe probably become lefs dangerous from the animal's licking the part. This is never done by rabbits and guineapigs, notwithftanding they are bit. I have affured myfelf of this particular in fuch a way, that I have not the fmalleft fufpicion of having been deceived.

More than two-thirds of both dogs and cats that I had bit in the nofe, never licked the part, although they could eafily have done it. I obferved them myfelf, and had them obferved, for whole hours. It is true, that thofe which bled a good deal licked themfelves if they could ; but it was evident, on obferving them, that they only endeavoured with the tongue, to free themfelves from the blood which tickled them in flowing down, and that this was no fooner effected, which happens in a moment, than they ceafed to do fo. — In the

expe-

experiments I made on dogs and cats, I prevented some of them, when they bled at the nose, from licking the part, and suffered others to do it. The disease was the same in all; and it is, therefore, certain, that simply licking the nose, whether in dog or cat, does not at all diminish the effects of the venom of the viper on that part.

CHAPTER VI.

Experiments on the Tendons.

SEVERAL modern physiologists have thought that the tendons are not endued with sensation. It is certain that it has not yet been proved clearly, that a tendon receives nerves, either from the muscle, or from the tunica vaginalis which covers it. Neither is it more apparent that it has blood-vessels, at least in any number, and sensible ones. It is therefore natural to suspect, that the bite of the viper on a tendon cannot be of any great consequence, and that the venom cannot act on this part. I wished nevertheless to consult experiment once more on this point.

In having the tendons bit by vipers, I was more than once on the point of being deceived; and if I
had

had not multiplied my experiments, and varied them
in feveral ways, as I did, I fhould certainly have
been fo. I fhall be circumftantial in relating fome
of the trials I made on the tendons, to fhow that it
is eafy for any one, even for an obferver, to be deceived, if he only follows fimple experiments, fince
the refult of them may vary, although there appears
to be no variety in the circumftances with which
they are made.

My experiments were made on rabbits, of which
I employed the largeft I could find, fome of them
weighing ten pounds and upwards.

Having removed the fkin from the *tendo achillis*
of a rabbit, and perfectly ftripped it of its tunick,
for a fpace of fix lines in length, I paffed under it
feveral folds of fine linen, to prevent the venom
from communicating to any other part. I wounded
the tendon in feveral places with a venomous tooth,
and afterwards covered it with bits of linen in fuch
a way, that it did not feem poffible for the poifon to
communicate to the neighbouring parts. The rabbit died at the end of thirty-fix hours. The tendon
was livid throughout its whole fubftance, but the
parts about it were not fenfibly changed.

I opened by an incifion, the fkin that covers the
tendines achillis of another rabbit, and ftripped the
tunick from both. The tendons were fmooth, fil-
ver-coloured, and free from veffels. I paffed feveral
folds of linen beneath them, and had them bit feveral times by two vipers, covering them with linen
in fuch a way, that the venom could not glide elfewhere.

where. The rabbit died at the end of thirty-eight hours. The blood in the auricles, in the ventricles, and in the large veffels of the lungs, was black and coagulated. There were feveral livid fpots in the lungs. The mufcles adjacent to the tendons were a little inflamed, and likewife had livid fpots in feveral places.

I repeated this experiment on two other rabbits, with pretty much the fame refult. Both of them were dead at the end of thirty-feven hours.

Although it clearly refults from the experiments I have juft related, that rabbits die afrer having been bit in the *tendo achillis* by vipers, I could not however conceive, that the death of thofe I have mentioned, was occafioned by the introduction of the venom, and its fubfequent difeafe.

It did not appear poffible to me, that a part endued with fo few vital principles as the tendon, which is not at all fenfible, and which may be cut both in men and animals with impunity, could be fufceptible of the action of a venom that has no influence either on the mouth or ftomach. I fufpected that thefe animals died from fome other caufe or circumftance I could not difcover.

In confequence of this fufpicion I determined to multiply my experiments, and to diverfify them as the cafe might require.

Having removed the fkin from the *tendo achillis* of a rabbit, and ftripped it of its tunick above and below, fo that it appeared fmooth and white, I wounded it with the point of a large and fharp needle,

dle, which pierced it through. The needle was well covered with venom, and I had put several folds of linen beneath the tendon. I wiped the tendon several times, removed the linen, and left the part expofed. I then introduced into the aperture made in the tendon, a bit of wood well covered with venom, and having withdrawn it, poured in a drop of pure venom. At the end of twenty-four hours, the tendon feemed difcoloured at the part wounded. The rabbit, however, ate conftantly, and at the end of fifteen days was recovered.

In another rabbit, I removed a large portion of the fkin that covers the joint of the knee, and ftripped the ligament that binds this part of the adipofe membrane. I wounded it obliquely with a venomous tooth, in eight places, at each of which a fmall drop of venom appeared. I made fmall incifions into each pun& ture, with the point of a lancet, which penetrated into the fubftance of the ligament without piercing it through, and conveyed the venom within. The rabbit recovered in eight days, and feemed not to have had any internal complaint. It ate conftantly, and continued lively and active.

Having ftripped the *tendo achillis* of another rabbit of its tunick, and put folds of linen beneath it as ufual, I had it bit feveral times by two vipers. I then covered it with linen, and removed that which was beneath. The rabbit for the firft few days feemed to have no complaint, but the wound in the tendon never healed perfectly. At the end of ten

days.

days its belly appeared to fwell, and, on its dying at
the end of teń days more, I found it to be drop-
fical.

These experiments feem to oppofe the former
ones, and to render it a matter of doubt whether
the bite of the viper on a tendon produces the difeafe of the venom, or not. The latter cafes feem
to indicate that it does not; but they are contradicted by the former ones. Now, as one of the principal refearches I propofed to myfelf to make, at fetting out on my experiments, was to difcover what
are the parts acted on by the venom of the viper,
and to obferve its different effects on the different
parts of an animal, I was determined to continue
my experiments on the tendons with a degree of
obftinacy, and to fee if I could fucceed in clearing
up this point.

Wifhing to obferve a greater degree of precifion
in my experiments, and fufpecting that the venom
might perhaps communicate to the neighbouring
parts in which the incifion had been made, and that
it might penetrate by degrees through the linen, how
much foever the latter might have been folded, I
conceived the idea of putting betwixt the folds, a
piece of thin and pliable lead.

Having ftripped the *tendo achillis* of a rabbit of
its tunick, I paffed beneath it a piece of linen folded eight times, in the middle of which I had put a
bit of lead, fuch as l have juft defcribed. I pricked
the tendon in feveral places with two venomous
teeth, and covered it in fuch a way that it was quite

4 .   enclofed,

enclofed, having a bit of lead both above and below. The animal died at the end of thirty-two hours. The tendon was black at the parts where it had been wounded, the mufcles near it were a little inflamed, and the blood about the heart in a diffolved ftate.

All thefe precautions, as we fee, could not prevent or retard the death of the animal. As this was, however, but one folitary cafe, I did not think it proper to ftop here.

I repeated the experiment on the *tendines achillis* of four other rabbits, well ftripped of their tunicks. I applied the linen and bits of lead as ufual, wounding the tendons with venomous teeth, that the venom might be more collected, and fpread as little as poffible on the tendon. In a word, I omitted nothing that could make thefe experiments decifive ones. The rabbits all died in lefs than forty hours. In fome of them the blood was coagulated about the heart, but not in others. The lungs were fpotted in all of them. The mufcles in the vicinity of the tendons were a little inflamed, and in two of the rabbits livid.

Thefe new trials did not clear up my doubts. If on one hand they rendered the action of the venom on the tendons probable, on the other hand I could not conceive how a part, that is neither fenfible, nervous, vafcular, nor mufcular, could either receive the difeafe of the venom, or communicate it to the animal, fo as to occafion its death. I reflected again, that I had employed large rabbits; that I had neither applied much venom, nor made ufe of

VOL. I.   Q   many

many vipers ; and that, on other occafions, I had
found a large rabbit to die late and with difficulty,
although it had been bit by feveral vipers, and died
with large wounds, and with the moft affured fymptoms of the difeafe of the venom. · This made me
fall upon a new fpecies of experiments, from which
I flattered myfelf that I fhould draw fome kind of
information.

I prepared the *tendo achillis* of a rabbit as above,
and paffed beneath it a piece of linen folded fixteen
times, with a bit of lead in the middle.. I pierced
the tendon in the ufual part with a venomous tooth,
and introduced a drop of venom collected at the
orifice, into the fubftance of the tendon, by a longitudinal incifion three lines in length, made with the
point of a penknife. The incifion did not penetrate
through. I left the tendon, venomed in this way,
during a fpace of fix or feven minutes, and then
foaked up the venom with dry lint, and by the means
of fmall pincers, wafhed the wounded part of the
tendon feveral fucceffive times. In proportion as
the linen became moift, I took hold of one end of
it, and drew it by degrees from under the tendon.
It was impoffible in this way for the water to foak
through the linen, and communicate the venom to
the adjacent parts. As I wafhed the tendon upwards of twenty times, it was not poffible for an
atom of venom to remain within it. The rabbit
died at the end of thirty-two hours ; the tendon was
almoft in its natural ftate, its colour being fcarcely
deepened at the part where the wound was made.

I re-

I repeated this experiment on two other rabbits, ufing the fame precautions. They were both dead in lefs than thirty-feven hours.

It occurred to me, that the linen left above and beneath the tendon till the death of the animal, might perhaps bring on fuch a change in the neighbouring parts, as to occafion a mortal difeafe.

Having removed the fkin from the *tendo achillis* of a rabbit, and ftripped it of its tunick, I put linen beneath it as ufual, and wounded it with a venomous tooth. I wiped the tendon with lint, and wafhed it a little, taking care that the water did not touch the adjacent parts. I then removed the linen, and applied frefh, to the part. The rabbit died at the end of thirty-fix hours. The parts about the tendon were in a natural ftate.

I prepared the Achilles' tendons of another rabbit in the ufual way, and wounded them with a venomous tooth. I left them in this ftate for two minutes, and then threw on them a great deal of water, repeating the ablution till I conceived that they were perfectly cleanfed in every part, and that the venom was either totally wafhed away, or diluted fo effectually, that it could not convey its action to the adjacent parts. I had found from former experiments, that when any other part of an animal was bit, or wounded by a venomous tooth, the throwing of any quantity of water on it was ineffectual, and did not prevent the animal from dying, and from having the ufual difeafe of the venom in

the

the part bit. The rabbit, the fubject of this experiment, died at the end of thirty-two hours.

Another rabbit treated in the fame way, not only
recovered, but feemed to have no other complaint
than that occafioned by the incifion of the fkin, and
other parts that cover the tendon.

Thefe refpective cafes, confidered circumftantially,
began to perfuade me, that the venom of the viper
is perfectly innocent to a tendon. To be certain of
this, I thought of varying my experiments ftill
more, and of making them in fuch a way, that they
fhould at length become decifive.

Having removed the fkin, and laid bare the *tendo
achillis* of a rabbit, I bound it very tight with a
piece of packthread, at both extremities of the tendinous fubftance. The ligatures were made in fuch
a way, that it was not poffible for any communication either of humours or fenfation betwixt the tendon and the animal to take place. I put the ufual
folded linen under the tendon, which I wounded in
feveral places with a venomous tooth, betwixt the
two ligatures. I covered the tendon with linen,
and the rabbit died at the end of thirty-two hours.

I repeated this experiment on another rabbit, the
tendons of which I tied in the way above, and had
it bit betwixt the two ligatures. I wafhed the
wounds with a great deal of water, which I threw
on with force, and then removed the linen from beneath. This rabbit died at the end of thirty hours.
Another rabbit died in twenty-feven hours, after
having been treated pretty much in the fame way
with

with the preceding one, with only this difference, that inftead of throwing a great deal of water on the tendons, I wafhed them by degrees, applying clean and dry linen, after removing that which I employed at firft.

It feems at length pretty clear, that the venom of the viper is not the caufe of the death of the rabbits, in the cafes in queftion, and that it has no action on the tendons. A doubt ftill remained, however, which it was neceffary to clear up. I had obferved that feveral mufcular fibres had found their way into the tendinous portions that form the *tendo achillis*, and conceived that the venom of the viper might firft communicate itfelf to them, and from them to the other parts of the animal. Notwithftanding there was but little probability in this conjecture, I wifhed to inform myfelf on the fubject by experiment.

Having removed a portion of the fkin from the *tendo achillis*, and ftripped it of its tunick, I deftroyed the mufcular fibres that defcend from the crural mufcles, and implant themfelves betwixt the three portions of this tendon. I paffed feveral doubles of linen betwixt thefe tendinous portions, in fuch a way that one of them was feparated from the other two, and enclofed in the linen. I wounded this portion with a venomous tooth, and covered it fo as to prevent the venom from touching any of the adjacent parts. The rabbit died at the end of thirty-two hours, with its heart and veffels filled with black and clotted blood.

Q 3   I re-

I repeated this experiment on the tendons of another rabbit, which died at the end of thirty-two hours. The wounded portions of the tendons were dark throughout their whole fubftance, and thofe which had not been wounded were ftill much more fo. The lungs were covered with livid fpots, and the heart and large veffels filled with black and clotted blood.

I made an experiment on another rabbit, in which, after deftroying the fibres betwixt the portions of the tendon, I paffed a folded linen under its whole fubftance, as I had done in the cafes related a little above, and wounded it, without feparating the parts, with a venomous tooth. The rabbit died at the end of thirty-three hours. The wounded tendon was become darker and redder in fome places, and the blood in the heart, and in the veffels that go out from it, was black, but fluid.

It appears ftill more, that the venom of the viper is not the caufe of the death of thefe animals, but that it depends on another caufe, probably on the denudation of the tendon itfelf. The following experiments remove all doubts on the fubject.

I got ready fix very large rabbits, all alike in fize, in two of which I laid the Achilles' tendons bare, as ufual, and wounded them with a venomous tooth, after having inclofed them well in linen. In two others I laid the tendons likewife bare, but pricked them with a needle in feveral places. In two others I fimply laid them bare, without wounding or pricking them.

I

I covered all the tendons in the fame way with linen. All the rabbits died; the two that had been venomed, died together at the end of thirty-two hours; of the two the tendons of which were pricked with a needle, one died in thirty hours, the other in thirty-two. The two in which the tendons were fimply laid bare, died, one in twenty-feven hours, the other in forty.

The inferences I have drawn from the experiments on the tendons hitherto related, are as follows :

I. That a tendon is not fufceptible of the difeafe of the venom.

II. That when a tendon is ftripped of its tunick, the animal almoft invariably dies, without any intervention of venom.

This laft inference is a very important one, and may be of fome ufe in the punctures of tendons in man. It fhows how dangerous it is to ftrip thefe parts of the *tunica vaginalis,* and how much this membrane ought to be fpared.

It remained for me to make one other experiment on a tendon, which I fhall relate here, and which may throw fome light on the nature and economy of tendinous fubftances, and of their nutrition. Having laid the *tendo achillis* of a rabbit perfectly bare, and deftroyed the mufcular fibres that enter into it, fo that there could be no longer any flefhy fibres or veffels in the tendon, I found that the rabbit ate a few hours after, and conjectured that it

Q 4 would

would therefore probably recover. In effect it lived, and at the end of thirty-four days recovered perfectly, the wound made in its skin healing up. I wished to see what had happened to the tendon, and whether, as one would have suppofed, it had dried up from a want of veffels. All the veffels about the tendon had been cut, and it was abfolutely feparated from every other fubftance throughout, except at its two extremities. I found it covered by a fubftance, partly fpongy or cellular, and partly callous, and fprinkled with feveral veffels. When I got to the tendon, I found it whitifh, fupple, and nourifhed, as ufual, although it did not any where appear to receive veffels.

Were repeated experiments fimilar to this one to be made, important confequences, and facts relative to the nutrition of certain parts, would perhaps refult from them.

The multiplied and varied experiments I made on the tendons, have been of very great ufe to me in the purfuit of my refearches. If any doubt had remained on the fubject; if I had not affured myfelf to a certainty that the bite of the viper on this part is not attended with any confequence; if I had apprehended that the venom could communicate itfelf to the animal through the medium of this fubftance, I fhould have had a thoufand doubts as to the parts on which the venom act, in an animal that has been bit. No fubject in nature is abfolutely indifferent; and when fuch rare and extraordinary effects are to be examined in the animal

mal body, nothing is to be neglected—nothing is
to be deemed unnecessary.

CHAPTER VII.

On the Nature of the Venom of the Viper. Descrip-
tion of certain Parts of the Head of the Viper, that
relate to the Venom.

BEFORE I examine the properties and nature of
the venom of the viper, I think it incumbent on me
to speak of some other particulars, that relate to the
canine teeth of this animal, to the bag or membrane with which they are naturally covered, and to
the veficle or receptacle of the venom, which the
moft modern writers continue to confound with the
bag or fheath of the teeth. I have treated of all
thefe particulars in the firft part of this work, but
think it effentially neceffary to introduce fome figures here, which will give a jufter conception of
what I have faid in the part alluded to, and of what
I fhall fay in the fequel.

I have judged it expedient to devote a chapter
entirely to this fubject, and to interrupt, as it were,
the chain of my experiments on the effect of this
poifon, applied to the different parts of animals; -
fince it is before all neceffary, that the reader fhould
know the nature of the venom, and not be left any
longer to bewilder himfelf in the erroneous opinions, and hypothefes deftitute of foundation, that
have been fpread by the writers who have employ-

ed

ed themselves on the occasion, both before and after the publication of my experiments. Too much cannot be said for this effect; for unfortunately, when the mind is prejudiced in favour of any opinion whatever, established by authority, and generally adopted, it seems to deny itself to even the evidence of fact.

In Mead's work on Poisons, a description is found of the head of the viper, the parts of which are represented by figures. These figures of Mead, or rather of Nicholls, who is the real authour of them, are so imperfect, that I have been obliged to substitute others I have had purposely made. I have found the former ones out of all truth and nature, and whoever will take the trouble to confront them with the parts from which they were drawn, will find no difficulty in agreeing with me.

Fig. 1. of Plate I. of this work (see the conclusion of the second volume) represents the two canine teeth of the viper of one side of the upper jaw, partly covered by a membrane in the form of a bag or sheath, open, as it is seen, to give passage to the teeth. Mead pourtrays this bag as if it was fringed at its edges. It is indeed sometimes found in this state, but is oftener without fringe or indentation, and such as I have represented it. The canine teeth are elevated and laid a little bare, as they appear when the viper is on the point of biting; when it depresses them, they enter entirely into the bag or sheath. It is easy to see, that if this bag were the receptacle of the venom, the latter would naturally flow out at the opening in it, and would

pass

pafs continually into the mouth of the viper. This errour is copied from Redi, who believed that the venom was contained in this fheath that covers the teeth, and that it was fecreted in a fmall gland feated under the eye.

Fig. 2. reprefents this bag or fheath, *s s*, opened with fciffars as far as its bafis, and likewife on the bone of the upper jaw. An elliptical hole, *n e*, with rounded edges, is feen at the bafis of each of the canine teeth, and a longer and narrower hole, towards the point of each tooth, *r a*.

At the fide of the teeth a bladder is found, *m*, refembling a fhepherd's purfe, which pierces the fheath by a long canal ending in a fmall orifice, *o*, betwixt the two teeth. The venom contained in the purfe or bladder paffes through this canal, and conveys itfelf to the tooth, entering at the hole fituated at its bafis, and going out by that at its point.

Fig. 3. reprefents the bladder or purfe feen with a lens. It is not formed of a fmooth even membrane, but is on the contrary full of plaits, as if it was a compages of inteftines, or of wrinkles and ridges. It is of a triangular fhape, and has a much greater width than depth. If it is cut tranfverfely, and examined with attention, it is found to be of a fpongy fubftance, and compofed of cells deeper than they are broad. Every thing concurs to the belief that it is not a fimple bladder or receptable of venom, but rather a true gland, very voluminous and of a particular ftructure, which feparates the venom

from

from the blood of the viper, and in which it is re-
ferved for the purpofes it is deftined to by nature,
undoubtedly for the animal's advantage.

The cellular ftructure of this fingular gland does
not permit the viper to exprefs with facility all the
venom it contains. I have found a difficulty in
forcing it out by a very ftrong preffure on the gland
with my fingers ; and indeed we have feen, that a
viper is capable of killing fix or feven pigeons, one
after the other.

The two figures, 4, 4, reprefent the receptacle of
venom in its natural fize, feen at its anteriour and at
its pofteriour part, and united with its excretory
canal.

Fig. 5, fhows a tranfverfe fection of the above, feparated by many fmall partitions, *s*, *c*, &c. and filled
with the venom which flows out drop by drop, as
at *r*, *a*, &c.   It appears in this way when obferved
with a lens.

Fig. 6, reprefents a canine tooth of a viper, with
all its internal cavities, and its two external openings.

s s, are the elliptical hole at the point of the
tooth.

c a, the opening of the hole at its bafis.

i i i, are the internal canal of the tooth, which
opens at the bafis *c a*, and at the point *s s*.

There is a large opening, *e*, which forms the bafis
of the tooth, and the fection of which is reprefented
by *m*.

r o, of the figure at the side, are the two openings,
e, of figure 6, which are difcovered by a fection of
the tooth, as at *a b*.

r, reprefents the fhape of the longitudinal hole of
the tooth.

o, reprefents the opening of the hole at the bafis.
This fecond canal of the tooth does not communicate with the firft, and only extends as far as *r*.

Fig. 7, reprefents two canine teeth on one fide,
having at their bafis feveral other teeth, more or lefs
formed, *a c r*. Thefe teeth are moft frequently
fix in number, and are fituated in the fheath, and
covered with a very fine cellular web, which binds
them, and unites them together. They are placed
one over the other, thofe that are uppermoft, or
neareft the canine teeth, being the largeft. The
others decreafe in proportion, and the two that are
neareft the canine teeth are perfectly alike in fize.
The points of all of them, even of the fmalleft, are
pretty hard, and well formed; they are channelled,
and end by the ufual hole at the point.

When thefe teeth are feven in number, the feventh is always the fmalleft of the whole. It is fituated below all the others, and in the middle. The
bafis of thefe teeth is not yet formed, and merely
confifts of a flexible, tranfparent, and whitifh jelly.
They are not only deficient at their bafis, but likewife want the oval hole;—the principles of it are
fometimes feen in the largeft of them.

Although the matter at the bafis of thefe teeth
appears a fimple jelly, even when it is viewed with
the common lens, the naturalifts would be very
much

much miftaken, if he fuppofed it to be non-organi-
cal. Stronger lens than the ordinary ones have
fhown me, that it is compofed of a very fine webbed
membrane, filled with extremely fmall round corpufcles. This membrane folds over itfelf, and
feems to fhow, even the holes, and the form that
the bafis of the tooth is one day to take. I have,
however, fometimes thought I could diftinguifh
this. Be that as it may, it is certain, that the gelatinous part of the tooth is organized, and that it
exifts in a ftate of organization a long time before
the tooth is entirely formed and in a perfect ftate.

*Of the Nature of the Venom of the Viper. It is ex-
amined as to its Acidity.*

The acquiring a perfect knowledge of the nature
of the viper's venom may be of the greateft importance to animal phyficks, and, at the fame time, very
ufeful to the human fpecies. Vague and fuperficial
notions on this point, have given birth to hypothefes, to theories, and laftly, to remedies.

The volatile alkali in a great meafures owes its
reputation, to the opinion of the venom of the viper
being acid.

The ancients were ignorant of what it confifted
in, and of the part of the animal in which it refided.
François Redi was the firft to eftablifh thefe points.
He found it to be a humour fimilar to the oil of
fweet almonds, which the viper conveys with its

tooth

tooth into the wound it makes in biting. But he was miftaken in almoft all he faid befides on the fubject of this venom. He believed that it refided in the bag, or plaited membrane, that covers the canine teeth. He could never difcover that it entered into the tooth itfelf, and flowed out of it; and thought that the fmall gland feated under the eye of the viper, ferved to fecrete this humour, into the nature of which I do not find that he ever made any refearch.

Before the time of Redi, there was none but very vague and confufed ideas on the venom of the viper. We owe to this celebrated Italian naturalift the firft advances into a fubject, which he found in its infant ftate, filled with hypothefes and vulgar errours. Thefe errours were proper to the time he lived in, and it required a genius like his to combat them, and to open a new road to truth. It feems as if we only throw off our ignorance to plunge ourfelves into errour, and that it is at this crifis that the man of genius gives us fome glimmerings of light. We fet out at ignorance, which leads us to errour, and from errour we at length arrive at truth. This is the ufual progrefs of human intelligence, and through thefe gradations the moft enlightened nations have paffed.

Mead is the firft who in any way examined the nature and qualities of the venom of the viper; but from a fatality to which even the moft diligent obferver is oftentimes fubject in his endeavours to make the earlieft opening to truth, Mead found

that

that this venom was acid, and that it changed the dye of the turnefol red, and even gave a reddifh tinge to the firop of violets.

A few years after, Mead himfelf, in a fecond edition of his work on poifons, retracted all he had advanced on the acidity of the venom of the viper, and confeffed, as a candid and ingenuous man, that it neither changes the dye of the turnefol, nor the firop of violets red, and that it is neither acid nor alkaline. Doctor James, who affures us that he repeated the experiments of Mead, has latterly found this venom to be acid; he however does not fpeak of the pofterior experiments of the above cited authour; neither does he inform us how, fuppofing him right at firft, he was deceived on the fecond occafion. This manner of publifhing ones ideas, or ones experiments, neceffarily tends to perpetuate doubts and hypothefes, fince, after all, the authority of one man is of as much weight as that of another, and fince we cannot guefs which of the two is in the wrong. Another writer, ftill more modern than Doctor James, has received it as a truth, that the venom of the viper is acid; fupporting his opinion on the bare authority of Mead, without telling us, that this authour has fince denied its acidity.

It was natural to conceive, that experiment itfelf had demonftrated to thefe writers, that Mead was miftaken the fecond time, and that he was right in his firft trials, when he found the venom acid; and this confideration obliged me to examine the

matter

matter afrefh. I hope that no doubt will any longer remain, and flatter myfelf that I have difco- vered the errour into which Mead fell when he firft examined this venom ; an errour againft which Doctor James was not able to guard.

I have fometimes, but rarely, found, that the venom of the viper gives the dye of the turnefol a light-red colour. This circumftance, inftead of inducing me to believe the venom acid, excited me rather to examine once more into the caufe of it, which might be accidental. I obferved that in thefe cafes it was not very pure, and on examining it with a microfcope, difcovered globules of blood floating in it. I then examined the mouth of the viper, and found the two bags or fheaths which cover the teeth flightly inflamed. It is not un- common to meet with vipers that are naturally in this ftate, and it is ftill more frequent to find thefe bags reddened after the vipers have bit. We like- wife frequently fee the venom ftained with blood, if its receptacle is too ftrongly compreffed. All thefe cafes may happen, and in all thefe cafes the dye of the turnefol may become red, without fup- pofing an acidity in the venom. It is, therefore, not unlikely that Doctor James has been deceived in the fame way with Mead. It is certain, that in the few cafes in which I have found the dye of the turnefol reddened, the venom was not pure, but was mixed with blood.

Aware of all thefe accidents, I took the utmoft precaution in collecting the venom. I generally cut off the head of the animal at a blow. Some hours after, when the mufcles had loft their motion, I opened the mouth carefully, and contrived that the points of the canine teeth fhould be free from their fheaths. I then made a gentle preffure on the receptacle of the venom, and received the latter in a glafs, as it flowed out at the point of the tooth. In this way it is ufually fo pure, that it appears, when viewed with a microfcope, like an oil, more or lefs yellow. No extraneous matter is obferved in it; and when I accidentally thought I perceived corpufcles floating in it, I did not employ it in the experiments of which I am about to give a detail.

When the venom was drawn in this way from the tooth, I never could perceive that it changed the dye of the turnefol red, however often I made the experiment, and I repeated it very many times. In moft cafes, I began with uniting a drop of venom with thirty drops of the dye or tincture, and not finding the colour of the latter to be changed, I added another drop of the venom, proceeding in this way to a tenth, or one-third the quantity of the tincture, which never either reddened or changed its colour, but only appeared not quite fo clear as before. I repeated this experiment too often to apprehend that I was miftaken.—I not only tried the venom with the tincture, or dye, of the turnefol, but likewife made the fame experiments on the blue juice of radifhes, a liquor very fenfible to the action of acids, even of the weakeft of them. It continued blue as before, without my being able to obferve

3   the

the flighteft change in it. I likewife had paper well tinged with the juice of radifhes, and let fall large drops of venom upon it;—the venom foon dried, and, except a yellowifh tinge it gave where it fell, I could perceive no change in the colour of the paper.

On feveral other occafions I diluted the venom with water, but could find no greater change in the paper on which I dropped it, than when I tried it pure.

I cannot deny but that I fometimes obferved a weak reddifh tinge on the blue paper, when I made the experiment in the following manner :—I covered a large ball of cotton with the paper, and on forcing the viper to make a ftrong bite at it, perceived this very pale tinge of red, at the parts the animal had pierced with its teeth. I did not, indeed, multiply my experiments fufficiently to be able to fay with certainty whence this tinge proceeded in thefe circumftances. We may fufpect that it was owing to a fmall quantity of blood from the mouth, blending itfelf with the venom; it is, however, very certain, that that taken from the veficle, neither changes the die of the turnefol, nor the juice of radifhes, red.

But even though we fhould agree that the venom of the viper may be capable of giving a red tinge to thefe liquors, does it follow of neceffity, that the volatile alkali. is a certain remedy againft this venom, and that the latter occafions death precifely becaufe it is acid?

R 2   The

The rock that men ufually fplit upon, a rock that the moft circumfpect philofophers have not always been able to avoid, is that it is fufficient for them to find a circumftance which accompanies the effect, to be too readily perfuaded that it is the caufe of it.

The innate defire we have of knowing every thing, makes us ftrive to explain every thing. If we fee an effect produced after the application of any given fubftance, we immediately endeavour to fee if there may not be fomething in that fubftance that may ferve in fome way to explain the effect; giving ourfelves very little trouble to examine whether the caufe we have difcovered is proportioned or not to the effect produced. This errour feems to have been committed by two men of the firft talents, Mead and Juffieu. Mead, when he publifhed the firft edition of his work on poifons, perfuaded of the acidity of the venom of the viper, judged that it muft neceffarily kill animals, becaufe it coagulates the blood as acids do.—Juffieu, perfuaded likewife of the acidity of the venom, from the authority of Mead, immediately found a fpecifick againft it in the volatile alkali *(a)*.

(*a*) Juffieu was not the firft after Mead to recommend the ufe of the volatile alkali ágainft the bite of the viper, but as he made a brilliant cure, it is to him that this remedy owes its greateft reputation.

The

The venom of the viper, as well as many other subftances, is formed of feveral principles we are ftill ignorant of. All the qualities we find in bodies do not conftitute their real nature;—fome of thefe qualities are accidental, others are not fo. The acidity, even though it fhould be conftantly obferved in the venom of the vipér, may, neverthelefs, be nothing more in it than an accidental quality; and the venom, in ceafing to be acid, may not ceafe to be a poifon. Chemiftry furnifhes us with a thoufand fimilar examples. It is, therefore, improper to deduce the caufe of the death from the acidity, and to deduce from the fame acidity the ufe of the volatile alkali, as a remedy; for even fuppofing the venom to be conftantly acid, and that this acidity cannot be feparated from it, does this enable us to fay, that it kills becaufe it is acid, and that the volatile alkali is its fpecifical remedy, becaufe it is capable of faturating it? The venom of the viper may likewife have feveral other qualities that we are unacquainted with, and may occafion death by each of them feparately, or by all of them together. Why then are we to fuppofe, that it derives its noxious qualities from its acidity? There are arguments that demonftrate the contrary.

Water abforbs about its own bulk of fixed air, and confequently a cubick inch of water can contain but very little more, if it does contain more, than a cubick inch of this air. It is not yet proved that a cubick inch of fixed air weighs an entire grain. A cubick inch of water weighs about 373 grains,

and confequently the fixed air contained in a cubick
inch of water, cannot be more in weight that its
373 part.   Now a cubick inch of water impregnated with fixed air, is capable of giving a red
tinge to 60 cubical inches of the tincture, or dye, of
the turnefol, that is to fay, to 22380 grains. Whence
we fee that the $\frac{1}{22380}$ part of a grain of fixed air
is capable of beftowing a fenfible tinge of red on
a grain of the dye of the turnefol.   Now granting
this hypothefis, there cannot be at moft in a grain
of venom more than the $\frac{1}{22380}$ part of acid matter,
and fince the thoufandth part only of a grain in
weight of the venom is capable of killing a fparrow, as will appear by and by, we muft fuppofe,
that the $\frac{1}{21380000}$ part of a grain of acid can kill
an animal fimply as an acid principle.

Who does not now fee, that even though it fhould
be granted that the venom of the viper gives a red
tinge to the dye of the turnefol, it would not, on
that account, follow, that it would kill becaufe
acid ? Its acidity would be fo inconfiderable, that
it would produce no fenfible change in the animal
body.   And where is that violent acid, or any other
principles of bodies, which is active to fuch a degree, that in diminifhing its quantity it does not at
length become innocent ?

Let any one fuppofe, if he will, that the acidity
of the venom of the viper is as great as that of the
glacial oil of vitriol (oil of vitriol concentrated to
the confiftence of *ice*) itfelf. If the mortal effects
of the former depended on its acidity, the glacial
vitri-

vitriolick acid, thrown on a wound, although in a very fmall quantity, would occafion the death of animals. Glacial oil of vitriol applied to a wound, may indeed render the ftate of it worfe, and may corrode the flefh, but will not kill the animal on which it is tried. Very little of it can be introduced into the circulation of animals, and the little that is introduced is then weakened by the blood with which it mixes. It is true, that, as well as the venom, it may kill, if injected in a fmall quantity; but this only happens becaufe it is not yet mixed with the humours, and weakened by them. Both the venom of the viper and the oil of vitriol may be abforbed by the veffels, and notwithftanding the former is abforbed in a very fmall quantity, and very much diluted by the blood, it will kill an animal which will not be killed by the oil of vitriol. The venom of the viper does not therefore occafion a very fudden death from its acidity, but from other principles as yet unknown to us.

Mead, who changed his opinion as to the acidity of the venom of the viper, never wavered however in his fentiments in regard to its fuppofed falts. He has always remained in the perfuafion of having obferved them floating in the yet fluid venom, foon after having taken it from the animal; and not only believes in the exiftence of thefe floating falts in the venom, but afferts that the venom itfelf changes to a fimple faline network, of a very beautiful ftructure, which he compares to a fpider's web. He fpeaks of the folidity and firmnefs of thefe falts,

R 4   which

which he defcribes minutely, and even gives a feparate drawing of them.   He adds, that he has difcovered here and there in thefe falts, fmall circu-
larly-formed knots, which are extremely folid, and
never lofe the fhape they have at firft taken.

This fubject, which appeared to me extremely
interefting, I examined very extenfively, in my
work publifhed in Italy, which forms the firft part
of this publication.   I even flattered myfelf, at
that time, that I had not only demonftrated the errour of Mead in an inconteftible way, but had likewife difcovered the fource of it.   To refute an errour in phyficks in a decifive manner, nothing can
be more effectual than the recurring to its origin.
But even this does not feem fatisfactory to certain
authours, who perfevere in maintaining, after the
authority of Mead, that the venom of the viper is a
mafs of falts; notwithftanding it is more than twelve
years fince Mead was refuted on this point.   I demonftrated at that time, that this venom is an homogeneous fluid, which, when taken pure from the
tooth, is never found mixed with falts floating in it,
nor with other heterogeneous particles ; and that
thefe floating corpufcles, when they are to be found,
are merely accidental, and are by no means falts. The
fmall knots feen by Mead, are nothing more than
fmall bubbles of air interfperfed in the venom.
Thefe fmall air-bubbles are never feen when the
venom is taken immediately from the veficle, and
may be made to appear at pleafure, by taking it
                                            blended

blended with the faliva of the animal, from the mouth of the viper *(a)*.

The faline net-work, which Mead fays he obferved, and which has been defcribed by many authours after him, is no other than the fragments of the dried venom, which, when taken from the tooth, and put on a bit of glafs, very foon dries, and whilft it is drying cracks in different parts, prefenting pieces and fragments very different from real falts. The Count de la Garaie made *falts* of the fame kind, by thoroughly drying his extracts on earthen plates, the glazing of which gave the hardened fragments a kind of fhining faline appearance.

If a drop of the venom of the viper, put on a bit of glafs, is examined with a microfcope, the fubftance of it will be feen to crack gradually at the circumference, where it dries fooneft. The fiffures in

---

(a) To have demonftrated the falfehood of any opinion whatever, is not a fufficient caufe for its being laid afide, if it is generally adopted by authours. Nothing lefs is needed for this effect than the renewal of the entire generation, to the end that it may flatter itfelf, that it cannot be reproached for rejecting an errour it has not committed. It required half a century to eftablifh the circulation of the blood, and the attraction of Newton, amongft philofophers. Man, always filled with a fecret pride, thinks that he is humbled if he difcovers himfelf liable to err ; and the vulgar, never to be trufted in their decifions, are of the fame opinion. We have unfortunately too many examples of this kind, not to perceive that the love of truth is by no means the firft fpring of human actions.

this

this part are fmaller and more crooked than elfewhere ; but, by continuing to obferve the venom,
larger, broader, and deeper ones, which advance
towards the centre of the drop where they end and
meet, are feen at every part of the circumference.
Thefe crooked lines are obferved very diftinctly
with a microfcope, running to the centre, and
lengthening in fuch a way that one might miftake
them at firft fight for fmall fnakes, writhing themfelves from the circumference of the venom to the
centre.   After all the fiffures are formed in this
way, they enlarge ftill more in proportion as the
venom becomes drier, and occupies a lefs fpace on
the glafs.

I do not know any microfcopical obfervation
more certain and more evident than this, and in
regard to which one may affure ones-felf with better grounds, that circumftances are thus, and not
otherwife.   But that not the fmalleft doubt may
remain, even in thofe who may not have an opportunity to repeat my experiments, I have thought it
incumbent on me to to reprefent, by feveral figures,
a drop of venom in the act of deficcation.   It will
be fufficient to give a glance at thefe figures to be
fatisfied of the truth.

Fig. 1. of plate II. reprefents a drop of venom at
the moment of its beginning to dry on a bit of
glafs.   The fiffures that are the moft curved, at the
circumference of the drop, are already entirely
formed, the venom beginning to dry at the circumference.   The others are feen becoming ftraighter,

length-

lengthening, and approaching to the centre, where the venom dries the floweft. When it is perfectly dry, the firft figure changes to the fecond, (fee figure 2.) in which the fiffures appear carried on to the centre, after having taken different curvatures. The fiffures in the centre are broader, becaufe the venom, which is there in a greater quantity, fepa-rates more on that account in drying.

Fig. 3. reprefents feveral fragments of the dried venom, in which the fiffures are defcribed by fpiral lines. Thefe fpires, as at *a*, are formed particularly, when the venom is dried in a confiderable quantity, and when it is pretty thick on the glafs of a watch. The fragments, which in this cafe are pretty large, open in the middle, and the opening, as I have juft faid, is of a fpiral form. The letter *e* reprefents a cleft that feparates the fragments from each other.

In Fig. 4. a drop of venom is reprefented, taken from the mouth of the viper, and dried. The fmall balls, or knots, of Mead are feen in it, as at *o*. Thefe fmall balls are real bubbles of air, which are made to difappear with the point of a needle, as all air-bubbles are that are produced in fluids. Letter *m* reprefents a cleft that feparates the fragments, as above.

It is an errour then, founded on ill-contrived experiments, that there are falts floating in the venom of the viper; and the regarding the fragments of this venom, when it is dried, as falts, is another errour. It is equal and homogeneous through-

throughout, and nothing of this can confequently be obferved in it.

Mead, who regarded the venom of the viper as a mafs of falts, likewife believed it to be cauftick and acrid when put on the tongue. He quotes himfelf and feveral of his friends, as having tafted it. He likewife obferves, that when the viper bites, and when the venom begins to find its way into the wound, the animal cries out, writhes itfelf, and and exhibits other manifeft figns of pain. Without pretending to decide at all on this queftion, which I have likewife examined in the firft part of this work, I fhall obferve here, that the experiment on dogs, which howl when they are bit, is not a certain and evident proof of the cauftick nature of the venom. Perhaps when it is united in thefe cafes with the fluids of the animal, it is decompofed, and acquires qualities it did not poffefs a moment before. It is true, that this howling which is mentioned, is fometimes obferved, but not always, and may be occafioned by its frequently happening that a nerve is pricked by the teeth of the viper, in which cafes the venom may caufe the fame pain as any other body, or fimple fluid, applied to the nerve itfelf.

If Mead tafted the venom, and found it cauftick, I have tafted it likewife, and have made others tafte it, and we have neither found it cauftick nor acrid. According to my fentiments, it has no kind of tafte when put on the tongue, and is neither perceived to fting or heat the part. It is

true

true that a fenfation is felt foon after, which may
have made thofe who believed it compofed of falts,
and who waited for fome extraordinary change,
fufpect that it was cauftick and hot. The fenfation it leaves when taken by the mouth, is that of a
torpor or ftupefaction in the part it touches. The
tongue particularly feems numbed; it even appears
to be grown larger, and its motions are flower, and
more difficult. This is certainly extraordinary,
but appears very different from the effects occafioned by cauftick and acrid fubftances, when put
on this part.—Laftly, Mr. Troja wifhed to tafte it
himfelf, and affured me that he found it neither hot
nor cauftick, but that this fenfation of torpor and
ftupefaction was the confequence of it in the
mouth. I can likewife take upon me to fay, that
I put five or fix drops at a time into the mouth of
fmall animals, fuch as rabbits, guineapigs, &c.
without even having been able to obferve any fwelling or rednefs. Thefe experiments, when made
on man, cannot be obferved without a degree of
repugnance, fince, after all, a fmall excoriation in
the mouth, or on the tongue, may caufe them to
be too dearly paid for by the obferver. I conceived that I could affure myfelf as to this particular in another way, and on a part even more fenfible than the tongue itfelf; that is to fay, on the
eyes of different animals.

I put fometimes one, and fometimes feveral
drops of venom on the eyes of a cat, and kept its
eyelids open by force. I let it fall into the eyes of
feveral

feveral rabbits without their perceiving it, and did
the fame thing to dogs.   It was feen running over
the tranfparent cornea and opake cornea, and getting within the eyelids.   I could not perceive in
any of thefe cafes that it acted as a cauftick or
acrid fubftance.

If Mead was miftaken when he believed the venom of the viper to be compofed of falts, he was
not miftaken, however, when he afferted that it
was neither acid nor alkaline, fince, in effect, it
neither effervefces with alkalies or acids.

It is needlefs, after the experiments recited in the
firft part of this work, to enter here into a detail of
thofe I was induced to repeat upon this occafion,
and which can no longer leave any doubt in the
minds of thofe who are fkilled in obferving.   It is
an eftablifhed truth that the venom of the viper
does not effervefce with any of the mineral or vegetable acids, nor with any kind of alkali we at prefent know of.   I have repeated thefe experiments
too often to have any doubt of having been mifled
by them.

But it is not fufficient to have fatisfied ourfelves
that the venom of the viper is neither acid nor alkaline; that it is not compofed of falts; and
that it is not corrofive to the palate; to inftruct
us in what it really is.   I do not know with what
other fubftance that is better known, it may be
made to agree.   It is principally to this point that
the efforts of obfervers fhould be directed, fince it
is certain that we are not thoroughly acquainted

with

with the true nature of any fubftance, although we
are more or lefs acquainted with the properties of
certain fubftances.

When the venom of the viper is yet liquid, it
unites in a greater or lefs degree with acids. But
we muft likewife examine it when dry.

I put feveral drops of very pure venom into the
concave part of the glafs of a watch ; as it dried,
it became yellow, and full of cracks. I poured oil
of vitriol on it, but no vifible folution followed. I
raifed from the bottom of the glafs, with a capillary
tube, feveral fragments of the venom, which floated
in the oil of vitriol without diffolving. At length,
after fome time, they feemed to begin to divide a
little, and though they were indeed reduced to a
kind of liquid pafte, ftill preferved their natural colour. There did not appear to be a true and perfect diffolution of them, at leaft during the time I
obferved them.

The marine acid, when poured on the dried venom, acts pretty much in the fame way as the oil of
vitriol. The fragments of venom do not appear,
in a ftrict fenfe, to be diffolved by this acid, although they are foftened by it.

The nitrous acid feems to have no greater power
to diffolve the dried fragments of venom, although
it at length foftens them. Notwithftanding the venom is rendered flexible by this acid, it ftill preferves a certain confiftence or tenacity which keeps
it together, and it becomes yellower. If examined

in

in this ftate, it appears to be compofed of an infinite number of very fmail fpherical corpufcles.

Thus then it appears that the ftrongeft acids have
but a very flow and weak action on the dried venom
of the viper, and that the diffolution they at length
occafion is but a very imperfect one.

Vegetable acids, however concentrated they may
be, do not diffolve this venom better than do the
mineral ones; and alkaline fubftances have no greater tendency to this effect.

I was likewife defirous of knowing whether effential oils would diffolve it, and on trying them did
not find them to poffefs that property.

The hepar fulphuris makes no greater impreffion
on it.

Thefe experiments, which I varied feveral ways,
made me fufpect by degrees, that the venom of the
viper might be either a gummy or a lymphatick
fubftance, feparated from the blood of the animal.
I had obferved a long time before, that the dried
venom appeared to be tenacious, like one of the
ftrongeft gums, when broke betwixt the teeth. Frefh
experiments were neceffary, however, to be certain
that it poffeffed the nature of a gum.

Chemifts know that gums neither diffolve in fpirit
of wine, nor in oil; but that they diffolve very readily in water. This kind of examination might
without doubt be fatisfactory, but it was firft neceffary to prove that it was not of the fame nature
as animal lymph, or the white of an egg. We
know that thefe fubftances coagulate in warm water,

ter, inftead of diffolving, as gums do.   I got ready
for this trial a great quantity of venom, which I
kept in a fmall capfular glafs till it became perfectly
dry.   On this venom I threw at once about half an
ounce of boiling water, by which it was inftantly
and effectually diffolved, inftead of being coagulated.   On repeating this experiment feveral times,
the confequence was invariably the fame.   The water, after having been thrown into the glafs, ftill
preferved upwards of fifty degrees of heat.

Having thus, by direct experiments, excluded
the hypothefis of a lymphatick animal matter, I
proceeded to the experiment of the fpirit of wine.

I had a good quantity of venom dried as ufual in
a fmall glafs, and poured on it half an ounce of highly
rectified fpirit of wine.   I left it in an undifturbed
ftate for upwards of two hours, when I found the
venom undiffolved at the bottom of the glafs.   I
broke it into feveral fmall bits with the fharp point
of a fmall glafs tube, and fhook the whole together
for fome time.   There was, however, no diffolution,
the fmall pieces of venom continuing whole, hard, and
of their ufual colour.   This experiment will always
fucceed in the fame way, if the fpirit of wine is
good ; but if it fhould contain too much phlegm,
the venom may be partly diffolved by it.   Even this
laft circumftance proves that the venom of the viper is a gummy fubftance, fince gums are very readily diffolved in water, which likewife diffolves the
dried venom, as I have affured myfelf an infinite
number of times.

If the venom is perfectly pure, the water does not lose any part of its tranfparence. Diftilled water is the beft calculated for thefe experiments.

I have frequently held the dried venom to the fire, and have increafed the heat by degrees, but it has never melted. If it is thrown on a live coal, it fwells and puffs up, but does not begin to take fire till it has affumed the appearance of a coal.

Another experiment now remained to be made, to render this matter decifive.

All chemifts know that gums diffolved in water are precipitated by fpirit of wine ; and that in this trial, the water in which they are diffolved becomes very white.

I put equal proportions of water into two fmall glaffes, and added to one of them a quantity of the venom of the viper, and to the other an equal quantity of gum arabick. The folution of gum arabick, which was made by heat, being reduced to the temperature of the liquor in the other glafs, I poured feveral drops of fpirit of wine into each of the glaffes. The number of drops thrown into each was pretty much the fame, when I began to perceive a whitifh cloudinefs, which difappeared a moment after in both folutions, at every drop of fpirit of wine poured into them. On continuing to throw an equal quantity of fpirit of wine into each glafs, I faw the white cloud, inftead of difappearing, extend itfelf over the fluids, which became whiter and more opake, at every addition of the fpirit. On ceafing to throw it in, I perceived that the white matter be-

gan

gan to precipitate; and on adding a few drops of the spirit afresh, found that there was no longer any separation in either of the liquors. At the end of twenty-four hours the precipitation was complete, and there was at the bottom of each glass, pretty nearly the same quantity of an equally white, soft, and paste-like, substance.

The venom of the viper, when dissolved in water and precipitated by spirit of wine into the form of a white powder or meal, cracks in different parts when dried afresh, and its fissures are of the usual reticular form.

When a clear and transparent oil of vitriol is mixed with the venom, precipitated by spirit of wine, and dried in a glass, it becomes at the end of a certain time, of a dark vinous colour. The same changes are observed in the solution of gum arabick in water, precipitated by spirit of wine. This gum, in drying, likewise adheres to the glass and cracks, and if a few drops of oil of vitriol are thrown on it, they become in the same space, of a dark vinous colour. The analogy betwixt the venom and the gum cannot be more perfect. They alike dissolve in water; they are precipitated in the same way by spirit of wine; the precipitated powder or meal is of the same colour; both of them crack in drying; oil of vitriol does not soften them till after some time; and changes its colour in the same way with each of these substances.

I made another experiment on the venom of the viper, which though it does not prove any thing es-

sential

fential as to the internal nature of this venom, is ftill
a further proof that it has a great analogy to the
gums.

I put fix grains of very pure dried venom into a
fmall matrafs, and added to it fifty drops of nitrous
acid, to throw off its airs.   There came off from it,
by the affiftance of heat, as much air, or perhaps
fomewhat more, as the matrafs could contain. This
was common air, a little changed in its qualities. I
continued the fire, and a clouded air came off, which
on examination I found to be compofed, one third
of fixed, and two thirds of phlogiftick air,

Gum arabick, in the fame circumftances, likewife
afforded fixed and phlogiftick air, and the confequences of both experiments were fo perfectly fimilar, that they might have been confounded together.   It is true that gum arabick likewife affords
nitrous air, but this only happens when it is in a
confiderable quantity.   If the quantity is very
fmall, the little nitrous air it furnifhes, decompofes
itfelf, and unites with the common air in the matrafs.

It feems then to be demonftrated, that the venom
is in reality a gum ; we at leaft fee, that it has all
the properties and principal characterifticks of fuch.

This venom is found in an animal, is elaborated in
its organs, and formed of its humours. It therefore ought to be confidered as a true animal gum,
particularly as the viper feeds on animals. Although we are unacquainted with any other animal
gum, I do not think that the venom on that account
                                                    fhould

should be denied to be such, since it has all the properties of a gum. It should therefore for the future, be inserted in the catalogue of gums, and this discovery may perhaps induce naturalists to examine, whether a gummy substance may not likewise be found in some other animal.

Allowing the venom of the viper to be a real gum, this will not lead us to conceive what it is that constitutes it a venom, since it is a known truth that gums are not so, and that they may be employed with impunity. It would be superfluous to relate the experiments I made on this subject, out of pure curiosity. I assured myself in a thousand ways, that gum arabick is entirely innocent when applied to wounds. But such is the condition of man, and such is the nature of what we call science. We at length arrive at certain bounds, to carry us beyond which all our efforts are to no purpose. The idea that the venom of the viper is a gum of some kind, does not serve in the least to explain to us, how this gum brings on a violent disease in an instant, and how it is that, in so small a quantity, it destroys life in so short a time. Whatever the principle that renders it venomous may be, the proportion of it is so small, that it does not at all change in it the usual properties of a gum; and the smallest vestige of this principle cannot be traced, whether the strongest microscopes are employed, or the venom observed in any other way. The most active substances are rendered such by quantities of matter that cannot be traced. The

S 3   point

point of a needle that has touched a variolow, puſ-
tule, preſerves its activity for years, and brings
about violent changes in the bodies of ſeveral perſons ſucceſſively pricked with it.

How far are we ſtill from penetrating the depths
of this myſtery ! Through how many difficult and
unknown ways muſt we not paſs, in getting ſome
inſight into a matter ſo obſcure and difficult as this !
Happy at length, if all the pains that are taken, if
all the efforts that are made, do not prove totally ineffectual.

This diſcovery, which enriches natural hiſtory
with a new gum, ought not to be neglected by naturaliſts. It may in time lead to a better knowledge
of the nature of the venom of the viper, and of the
complicated effects it produces. It may perhaps be
one day uſeful to us in enabling us to comprehend ;
why animals with cold blood are ſo long in dying
of the bite of the viper ; why there are ſome that
are not killed by it ; and why the venom, in whatever way it is introduced into its body, is altogether
innocent to the viper itſelf. If the animals with
cold blood that die late ; if the others that do not
die ; if the viper to which the venom is not at all
hurtful ; had humours or parts of ſuch a nature, that
they could be but little, or ſlowly, or not at all
changed by this animal gum : we might then in
ſome way explain a ſubject which is as yet very obſcure, and which does not ſeem capable of being
cleared up, till after we have acquired a thorough
knowledge of the venom itſelf, and of the moſt latent

tent principles and qualities of the animal bodies on which it acts.

On Bees, Drones, and Wasps.

In the first part of this work, I related a few experiments on the venom of the scorpion, and on the humour which flows from bees when they wound with their sting. I have had occasion since to make some other observations, not only on bees, but likewise on wasps, hornets, and drones. I do not know that any naturalist has examined in a proper manner, the liquor with which these animals are provided, that wound with a sting. Indeed Mead says that he found the humour of bees, to be composed of very small saline needles, or points. He assures us that he examined it with a microscope, and found it filled with these pointed salts. I do not know whether this observation made by Mead, has been confirmed or not by other naturalists; but can for my own part take upon me to say, that I never have been able to find any thing saline in this humour; whatever attention I paid in investigating it; and notwithstanding I employed the strongest lens for that purpose. I am persuaded that Mead has been mistaken in this particular, as he was in observing the venom of the viper. He assuredly saw particles floating in this humour before

it

it was dry, and immediately perfuaded himfelf that they could be no other than floating points.

We may eafily conceive that Mead only exami-ned this humour, in an impure ftate, and mixed with corpufcles that were foreign to it ; and that this was fufficient to induce him to believe it com-pofed of falts. He was deceived on this occafion, as he was in his opinion of the venom of the viper, in which there is nothing to be met with of all that he fancied he faw; and he feems in both cafes to have erred exactly in the fame way. The humour of bees, after the manner of the venom of the viper, cracks in drying, and prefents the ufual fharp and regular fragments. This was fufficient to perfuade Mead that it was a true falt.

I can venture to fay, that when the obfervation is well made, nothing can lead to fuch an opinion. But if, in expreffing the liquor from the bee's fting, the greateft care is not taken to prevent the break-ing and mixing any thing with it, it may eafily be charged with other irregular bodies ; and when it is put on the port-object, fome fmall degree of mo-tion may likewife be obferved in thefe bodies, which may float in a greater or lefs quantity. But this accidental motion, which is not proper to thefe fubftances, foon ceafes altogether, when the humour is left undifturbed. By degrees it dries, and, in drying, breaks, cracks, and forms angles and points.

When the venom of the viper and the humour of bees, are dried and obferved with a microfcope, no fenfible difference can be obferved betwixt them.

l have

I have only taken notice, that the humour of bees, expofed to the open air on a bit of glafs, is much longer in drying than the venom of the viper, and that the cracks or fiffures in the former, are likewife formed much later than thofe in the latter, fuppofing the degree of deficcation in the two fluids alike.

Thefe two humours not only agree in the appearances their parts prefent in drying, but likewife in other qualities. If a bit of the dried humour of bees is ftrongly compreffed betwixt the teeth, it, as it were, glues them faft together; and exactly the fame thing happens on trying the venom of the viper, and all hardened gummy fubftances. The dried humour of bees likewife diffolves in fimple water, and refifts the action of fpirit of wine, as the venom of the viper and gums in general do; fo that I am almoft inclined to believe that, as the venom of the viper is moft affuredly a gummy fubftance, this humour is fo too. Indeed the quantity one is able to collect of it is fo very fmall, that one can fcarcely attempt to make any certain experiments on this fubftance; the confequences, however, of thofe I have made have been fufficiently uniform to lead me to think that I cannot eafily have been miftaken in what I have conjectured.

I have met with the fame fuccefs in examining the humour of wafps and drones, and of the other flying infects in general, that wound with a fting, and are provided with a humour. In all thefe, the humour is bitter and acrid, and has all the appear-

3   ance

ance of being of a gummy nature. When left to dry on a bit of glafs, it cracks throughout like the venom of the viper, and when chewed, is tenacious, glutinous, and elaftick.

But it muft not therefore be thought to be the fame as the venom of the viper, and that it has all the other qualities of this poifon. The venom of the viper neither has any fenfible tafte when taken into the mouth, nor is fufficiently acid to give a red tinge to the tincture of turnefol, or juice of radifhes. The humour of bees, and of the other analogous infects, the moment it is applied to a piece of paper that has been previoufly ftained with the juice of radifhes, gives it a flight red tinge, which afterwards changes to a pale yellow, fo that one would conjecture that this humour deftroys the blue colour of the paper. This experiment, which has been repeated feveral times, and always attended with the fame fuccefs, proves that this humour is united with an acid, and not with an alkaline principle; we fee, however, at the fame time, that the quantity of acid it contains is very fmall, and abfolutely incapable, as an acid principle, of occafioning the fmalleft fenfation on the tongue, or in the part pricked by the fting of the animal.

A quantity of water impregnated with an equal bulk of fixed air, gives a red tinge to paper ftained with the juice of radifhes. This tinge, which is pretty ftrong, continues a confiderable time. A fmall quantity of water impregnated with fixed air, fcarcely contains a fufficient degree of acid to be fen-

4   fible

fible to the tafte, and is likewife entirely innocent
when applied to wounds.

We muft therefore regard the hypothefis of thofe
naturalifts, who have advanced that this humour
occafions a fwelling in the parts into which it is
introduced, and that the volatile alkali, as faturating
the acid principle, is a remedy againft it, as falfe
and erroneous.

Experiment feems to indicate that this humour
acts by the medium of a bitter and cauftick principle, which is neither acid nor alkaline. If it is put
on the tongue, it has a hot bitter tafte, as I obferved
before, and not that of an acid or alkaline fuftance.

There are many fubftances which, without being either acid or alkaline, are hot and acrid to the
palate, and are productive of violent and difagreeable fenfations. Cantharides, and feveral aromatick
plants, are of this clafs. In the prefent cafe it appears certain, that neither the pain, (which is frequently infupportable, and greater than that which
would be caufed by oil of vitriol itfelf) nor the fwelling nor inflammation of the parts, can be brought
on by an acid principle introduced into the fkin of
the animals that have been ftung ; and therefore the
theory laid down by certain authors to explain the
effects of this humour, muft be regarded as abfolutely falfe, and the confequences they have deduced from it as no truer than the theory itfelf. A
pretended concentrated acid, a *naked* acid, an un-
combined acid, and a phofphorick acid that produces fuch wonderful effects, are hypothefes that
                                                    are

are not capable of refifting the inveftigation of rea‑
fon and experiment, and are unworthy the enlight‑
ened age we live in.   It is no longer the feafon to
imagine nature; we muft confult her.   If che‑
miftry has increafed the number of our intelligences,
the abufe of chemiftry has frequently retarded our
progrefs in the fciences.   It has frequently led us
into errour, and has fubftituted hypothefes for
facts and experiments.

Although bees, and the other infects that are an‑
alogous to them as far as relates to the humour
they throw out at their fting, are not capable of
killing, I think notwithftanding, that if they are
not confidered as venomous animals in the moft re‑
ceived fenfe, they fhould at leaft be confidered as
animals that fecrete in their bodies a fmall quantity
of a matter, which is not deftructive fimply becaufe
it is in too fmall a quantity.   The moft active poi‑
fons and venoms, fuch as arfenick, corrofive fub‑
limate, and the venom of the viper, when taken or
applied in a very fmall quantity, not only do not
occafion death, but do not even produce a fenfible
derangement, very far from their effects equalling
thofe that are produced by a large hornet, when it
wounds with its fting.   Thefe quantities, how‑
ever, although very fmall, are capable of killing the
fmaller fpecies' of animals, whilft more confiderable
ones are not fufficient to kill thofe of the larger fpe‑
cies'.   Hence we fee, that the difference entirely
confifts in the quantity of the venom, and in the dif‑
ferent degrees of ftrength in the animal that receives

it, and not in the nature of the venom, which is always the same. The venom, for so I shall call it, of bees, is very active, considering the smalness of its quantity, and we may easily judge from the pain and inflammation it excites in an instant, that if the dose of it were increased, it would produce the most violent derangements, and perhaps even a very speedy death. Nay, I am almost inclined to think, that a grain in weight would kill a pigeon in a few seconds. The difference that is found betwixt the sting occasioned by a bee, and that of a hornet, notwithstanding the difference in the respective quantities of their venom is but very inconsiderable, is very great. The same thing may be observed of the common scorpions of Italy, and those of other countries, as well as of the bite of spiders. The larger produce in general the greatest derangement, and those of Africa, or of Asia, even occasion death : all of them, down to the smallest, possess a greater or less degree of activity.

There are other animals, particularly insects, which when they bite or sting, bring on a very violent pain and inflammation, so that they may reasonably be suspected of introducing a caustick and venomous humour into the wound. In this number we may reckon ants, which insinuate into the small wound they make in biting, a very sharp and poignant humour, which they force from a vesicle seated in the hinder part of their body. I shall not make a digression here to speak particularly of this humour, because I have treated of it in a very ample

ple

ple way, in a paper of which the object was an examination of the *acids of animals*, &c. and particularly of *the nature of that of ants*, printed in the jour-
nal of the Abbé Rofier. I there demonftrated that
the humour of ants is a true acid, and that it is in
reality the acid of fixed concentrated air, deprived
of its elafticity, and rendered liquid.

PART III.

CHAPTER. I.

Action of the Venom of the Viper on Parts of an Ani-
mal that have been previously bit.

THE subject of this part is the most interesting
one that the matter it treats of can present to a phi-
losophick observer.

All the questions that are here discussed become
of consequence, since they tend to throw great
lights on the nature of venom. The animal economy itself is by their means better explained, and
many hypotheses that have been imagined, fall before experiment. It is the touchstone that makes
us soon distinguish all that does not belong to nature, all that is the effect of art, of prejudices, and of
the imagination; in a word, of man.

Experiment alone may conduct us through the
unknown paths of nature, and may lead us to new
and unexpected truths. But at the very time that
man, profiting by this torch, is making bold strides
towards the truth, and soars as if he meant to go-
vern nature herself, she stops him every moment,

ment, and by only difcovering herfelf to him in
part, feems afraid of being recollected; fhe thus
continually reminds him of his weaknefs, and fhows
him that his hopes are either vain, or confined within very narrow limits.

Man, who affigns to comets the courfe they are to
keep, and who fixes the time that is employed by
the light in its progrefs from the fun to our hemi-
fphere, is not, with all this knowledge, acquainted
with the air that furrounds him, or with the fire
that warms him. Such is our condition, and fuch
is the ftate of human fcience.

The firft queftion that prefents itfelf, after what
has hitherto been related, is to know whether the
venom of the viper is a poifon to all the animals
with warm blood. It will be feen in a little time,
that this large body of animals has not been feparated without defign from the other, which com-
prehends thofe that have the blood cold. When I
fay that a fubftance is venomous to an animal, I
mean to exprefs, that it produces in it very violent
diforders, although it is only introduced into its body
in a fmall quantity.

To reply properly to the queftion I have juft propofed, it is certain that all the animals with warm
blood exifting on the habitable globe, fhould be bit
by vipers. The fubject is not fufficiently interefting to deferve fo long and difficult a labour. However, if the analogy betwixt the different animals
with warm blood may be allowed, I am not afraid
to advance, that the venom of the viper is a poifon

to

to all of them. We have feen that it has proved fo to all the feven fpecies' that have hitherto been examined; and I very well recollect that I could not find any animal in Italy, with warm blood, to which the venom of the viper did not prove a real poifon. I tried it on all the birds I could meet with, and on all the quadrupeds I could procure, provided they were of a moderate fize; as to the horfe, the camel, and the ox, fetting afide their bulk, I could not procure them eafily for this purpofe.

We may therefore, I think, conclude with a great deal of reafon, that the venom of the viper is a poifon to all the animals with warm blood; that is to fay, that neither of them is beyond the reach of the effects it ufually produces, when it is introduced into the body in a fufficient quantity.

The fecond enquiry, which fprings immediately from the firft, is to know whether the venom of the viper is a poifon to all the animals with cold blood.

It has already been feen in a former part of this work, that even the frog, a cold animal, and one very hard to kill, dies in a few hours, if it is bit by the viper. This, however, is not fufficient to admit a certain conclufion, that all the other animals with cold blood would die in the fame way. We frequently incur the rifk of being deceived by this method of employing analogies on too narrow and limited a fcale.

A fingle fpecies of animals is not fufficient to furnifh an analogical argument of any weight.

Had five or six hundred kinds of animals with cold blood been examined, and had certain symptoms of poison been observed in all of them after they had been bit, the analogy, in this case, would have formed an argument of probability, and we might have been enabled to draw conclusions on this subject, not only in regard to animals with warm blood, but likewise as to those that have the blood cold.

We can scarcely do otherwise than suspect that the venom of the viper is innocent to the viper itself. This animal, in all the diseases or wounds of its mouth, would otherwise run a very great risk of killing itself with its own venom. It is not very unusual to find vipers with the bag or sheath of their teeth inflamed and bloody. Small red spots are frequently observed in the mouth of this animal when it bites, and it is besides easy to conceive, that if it should be bit in the mouth by any other animal, its own venom would prove destructive to it, if its particular nature did not guard it against such an accident.

The venom of the viper is constantly secreted and laid up, in the spongy gland. This gland has its canal continually open, through which the superfluous venom that cannot be contained in the gland, is forced to shed itself into the viper's mouth.

However, it is easy to have recourse to experiment. In the first part of this work a detail may be found of a great number of trials I made on this subject, and from which it results, that the venom

of

of the viper is not a poifon to vipers, but that, on
the other hand, it is altogether innocent to them.
I was defirous of repeating feveral of thefe experiments over again, and out of the great number,
which brevity obliges me to omit, I think it fufficient to relate a fingle one. '

After having enraged a viper very much, I forced
it to bite itfelf feveral times in the part towards its
tail; it, however fuffered nothing from this, although it had certainly forced its teeth well into
the part. I repeated this experiment on three
other vipers, with the fame fuccefs. It is therefore very certain, that the venom, or bite, of the
viper, is entirely innocent to this animal when it
bites itfelf, and it likewife is when one viper bites
another.

But this very fingular exception is not confined
to the viper. There are other animals to which
this venom is innocent, and others again in which,
although they were fmall, one or two vipers are
fcarcely capable of producing any fenfible change.
I have mentioned fome of thefe *cold animals (a)* in
the firft part of this treatife, but to come at the
number of them, the experiments fhould be extended to other fpecies' I could not at that time
procure, and on which I thought it fuperfluous to
make this trial.

If it is altogether extraordinary, that the fame
matter is entirely innocent to feveral fpecies' of

_(a)_ Animals with cold blood.

animals, and that it is mortal to an infinity of others, it is much more furprifing, and at the fame time more difficult to conceive, how, and by what principles, it happens, that an infipid gum, as far as we can perceive, excites the moft violent diforders in fo many very large animals, and that it does not bring about the fmalleft change in others that are incomparably fmaller and weaker.

The known diftinction of animals with cold, and animals with warm blood, which is only founded on a greater or lefs degree of heat, and on fome other trivial difference in the circulation of humours, is of no ufe in the prefent cafe, fince there are certain animals with cold blood that die of the venom, and others again that are not at all acted on by it.

If a comparifon is formed betwixt two cold animals, one that dies of the difeafe of the venom, and the other that furvives its action, they will be found to poffefs the fame organs, the fame circulation, an equal tenacioufnefs of life, and, in a word, to the eyes of the obferver, they will both of them appear perfectly alike.

What is it, then, that caufes this matter which flows from the viper's tooth to be a poifon to one, and not to the other? We are not only entirely ignorant of this, but it appears that we are likely always to remain fo. To obtain fuch a knowledge, it would be neceffary to be acquainted with the moft hidden nature of this extraordinary animal gum. It would be neceffary to penetrate into the moft internal and latent fubftance of the folids and

fluids

fluids of animals with cold blood, to know the mechanism of their organization, and to comprehend
perfectly the principle of life. We might then reply to all that could be asked on this head. But
how is it possible to acquire so extensive an information, whilst the activity and penetrability of our
organs are so limited and confined?

But if we are not permitted to know what this
very active principle of the venom of the viper is,
which when it is introduced into a living animal
causes its death; we are allowed, however, to enquire into the quantity of this venom, that is neceffary to kill an animal of a certain size. This enquiry, very curious in itself, cannot but be of some
use in the practice of medicine, particularly in cau-
tioning us against thinking the danger greater than
it really is, when any one of our own species has the
misfortune to be bit by this animal.

To be able to speak with some degree of precifion, in this research, it was proper to begin by determining very small quantities of venom, and by
introducing them without lofs into the substance of
the body of a living animal. It was likewise expedient to operate on very small animals, that would
die soon and to a certainty, to the end that the confequences might be lefs equivocal. It is true, that
by an endlefs multiplication of experiments, the
same consequences might at length be obtained from
large animals; but a longer time, and greater con-
veniences would be required, and one ought besides

to

to be perfuaded of the importance of the undertaking.

In the following experiments, I made choice of
fparrows and young pigeons, knowing them by experience to be readily killed by the venom.

To determine fmall known quantities of venom,
I began by taking four grains in weight of the venom of the viper, and mixing with it eight grains
of diftilled water. I then, with a fmall brufh, fpread
it equally over a fquare inch of thin paper. This
may be done with a fufficient degree of eafe and
precifion to exclude any confiderable errour, and
indeed I found that the halves and quarters of the
fquare inch of paper were of the fame weight when
dried.

I cut this paper in two, and again divided one of
the halves, continuing in this way till I had made
fix divifions, reckoning the firft. I then did the
fame with the other half, that I might have two
pieces of the fame fize, and of each fize, inftead of
one.

I ftripped the mufcles of the leg, in ten fparrows, of the fkin, and bound upon them the ten bits
of paper I have mentioned. The confequences, be-
ginning with the larger bits of paper, $\frac{1}{4}$, $\frac{1}{8}$, $\frac{1}{16}$, $\frac{1}{32}$,
$\frac{1}{64}$, were as follows. Of the two fparrows to
which the papers marked $\frac{1}{4}$ were applied, one died
at the end of fifteen minutes, the other not till the
end of thirty-five. One of thofe with the papers
marked $\frac{1}{8}$ died at the end of an hour, the other furvived. One of thofe with the papers marked $\frac{1}{16}$
died

died at the end of two hours, the other recovered. One of the two with the papers marked $\frac{1}{32}$ died at the end of two hours, the other at the end of five. And of the two with thofe marked $\frac{1}{64}$, one died at the end of three hours, the other at the end of feven minutes.

On repeating this experiment, the confequences were ftill more irregular. I therefore abandoned this method, as altogether infufficient and delufory. This is probably owing to the paper, which, when put in contact with the humours of the animal, may not allow itfelf to be either equally or entirely deprived of the venom that adheres to it. This obliged me to have recourfe to another method, which is perhaps lefs exact in determining the precife quantity of venom, but which has afforded me confequences as conftant and as uniform as can be expected in fo difficult an undertaking.

This is the method I employed :—I took a given quantity of venom, for example three grains, and fpread it over a bit of glafs, in fuch a way that it occupied a determinate fpace of a circular form.

The venom in the centre was not more than a quarter of a line in depth.

I procured a fmall capillary glafs tube, which terminated in a fmall fcoop of about half a line in diameter. I plunged this fmall fcoop vertically into the centre of the venom, and drew it out in the fame direction.

To determine the quantity of venom that adhered to the fmall fcoop, and to know at the fame time,

T 4 whether

whether this quantity would be conftantly the famo, I put the three grains of venom, laid on the bit of glafs, into a very nice ballance, and plunged the fmall glafs fcoop ten times fucceffively into the liquor, taking care to wipe the fcoop well every time. After the ten plunges, I found the equilibrium to be loft, and that about the $\frac{1}{100}$ part of a grain of venom was deficient. I now plunged the fmall glafs fcoop ten other times fucceffively into the venom, and the ballance having again loft its equilibrium, found that the venom was diminifhed about the $\frac{2}{100}$ part of a grain. With a little practice, one can make this experiment in lefs than two minutes, and in that fpace the three grains of venom are not fenfibly diminifhed in weight by the natural evaporation, as I have affured myfelf by trial. I cannot take upon myfelf to fay, that all the quantities are here rigoroufly the fame; nay I agree that, in repeating this experiment feveral times, a fenfible difference, which I have indeed met with myfelf, cannot fail to occur: but all thefe differences taken together can make but a very trifling variation in the quantity of venom that adheres each time to the fmall fcoop. On the whole, I can lay it down as an eftablifhed rule, that the fmall glafs fcoop plunged perpendicularly into the venom, in the way I have defcribed, carries away with it about the $\frac{1}{1000}$ part of a grain of the venom of the viper.

I laid bare a portion of the right leg of a fparrow, and made a fmall longitudinal incifion into the mufcles with a lancet. Into this incifion I introduced,

at

at the very inftant, the fmall fcoop armed with venom, and kept it in this fituation for thirty feconds.
The fparrow died at the end of two hours, with a
lividnefs of its leg.

I repeated this experiment on fix other fparrows,
exactly obferving the fame circumftances. They
all died, one after the other, at the end of the following times, expreffed in hours, 2, 2, 3, 4, 5, 7.

I again repeated it in the fame way on twelve other
fparrows, and the confequences were ftill more irregular than thofe of the preceding experiment.
One of the fparrows died at the end of four minutes, another at the end of three days, and another
at the end of five. The fymptoms of the difeafe
were, notwithftanding, indubitable in all the three.
The nine others all died at the end of the times expreffed by the following numbers, which denote fo
many hours ; that is to fay, 2, 3, 3, 5, 6, 9, 10, 12.

The firft confequences fhow, that the quantity of
venom I have mentioned is fufficient to kill an animal of the fize of a fparrow, but that it produces in
thefe animals very unequal effects, and a difeafe of
greater or lefs violence. An animal that dies at the
end of three minutes, and another perfectly fimilar
to the firft, that does not die till the end of five days,
prove that the difeafe in each of them has been very
diftinct. But fuppofing the quantities of venom
which were introduced to have been equal, and that
the incifions were fo too, a little more, or a little
lefs, blood, oozing from the incifed veffels, might
have caufed all this difference, fince it might have

occa-

occafioned a greater or lefs quantity of venom to enter into the circulation of the humours, or, to exprefs it ftill better, into the animal itfelf.

I wifhed to fee whether I could bring on a more fpeedy death by doubling the quantity of venom ; and being at a lofs for a certain method of collecting this double quantity together, I made two incifions inftead of one, and introduced the fcoop I have mentioned into each. Twelve fparrows on which I made this experiment, all died, but at very different intervals. One died at the end of three minutes, another at the end of twenty-feven, and a third at the end of forty ; the others at the end of the hours expreffed by the following numbers, 1, 1, 2, 2, 2, 3, 3, 5, 6.

The experiments I made on fparrows, and the method of introducing a given quantity of venom into the incifion in their mufcles, have occafioned me to make a very interefting obfervation. I ufually kept the fmall fcoop in the incifion for about twenty feconds, and obferved that the lips of the wound became livid at the end of a certain time. I regarded this fymptom as a fure mark of the communication of the difeafe, and it will be feen by and by that I was not miftaken in this conjecture.

I was defirous of feeing what effect would refult from a certain quantity of venom, applied to an animal larger than a fparrow.

I ftripped a part of a pigeon's leg of the fkin, in fuch a way that the mufcles beneath were entirely bare. Having made the ufual incifion, I introduced

duced the fmall fcoop, which had beed firft plunged into the venom, and kept it there till I faw the lips of the incifed part become livid; this happened, as in the fparrows, in the fpace of about twenty fe-conds. At the end of half an hour the leg became flightly livid, but neither feemed fwelled nor hard. The pigeon neither died, nor fuffered in any fenfi-ble degree.

I repeated this experiment on fix other pigeons, exactly obferving the fame circumftances. One of them had not the fmalleft fymptom of difeafe, nei-ther did the incifion become livid, notwithftanding I kept the fmall fcoop in it for upwards of a minute. Four others had fymptoms of the difeafe of the ve-nom, and two of thefe did not recover till after forty hours had elapfed. The fixth, like the firft, was not at all difeafed; the incifion, however, I made in its leg, bled at the time I introduced the venom.

I repeated this experiment, with the fame cir-cumftances, on fix other pigeons. One of them died at the end of fix hours. Three had all the fymptoms of the difeafe of the venom, and did not recover till the third day. Two others had not any fymptom of the difeafe. I think it proper to obferve here, that the incifions in the leg, in thefe two laft, bled very fenfibly when I introduced the venom. This fhows that the blood which flows from the veffels, may prevent the venom from en-tering them, or from remaining in them after it has entered,

5.   I re

I repeated this experiment over again on twelve pigeons, one of which died at the end of ten hours. Two others were exceedingly ill. The other nine had no fenfible complaint.

Thefe new experiments fhow, that the quantity of venom which ufually kills a fparrow is not fatal to a pigeon. We fee, however, at the fame time, that a cafe may occur, in which fuch a quantity of venom is introduced as is capable of killing a pigeon, notwithftanding that the quantity employed in the experiment would fcarcely have been fufficient, generally fpeaking, to kill a fparrow.

I was defirous of trying on pigeons, as I had done on fparrows, what would be the effect of two incifions, and a double quantity of venom.

Having laid bare the mufcles of a pigeon's leg, as ufual, I made two fmall incifions into them, introducing into each, the fmall fcoop armed with venom, in the accuftomed way. The livid fpot appeared at each incifion, and almoft the whole of the leg became livid, and remained in that ftate for two days, at the end of which time the animal was perfectly recovered.

I repeated this experiment on twelve pigeons, and the confequences were fomewhat various. Two of the pigeons died at the end of three days. The others had all a lividnefs of their legs, and all recovered notwithftanding.

On repeating this experiment on twelve other pigeons, four of them died; one at the end of fix hours; another at the end of twenty; and the two
others

others not till the fifth day.   All the others had
the difeafe of the venom, but recovered.

Inftead of making two incifions only, I tried the
experiment on twelve other pigeons, by making
four incifions at the fide of each other.   Nine of
thefe pigeons died ; one in ten minutes, two in an
hour, two in two hours, and three in five hours.
The other three had the difeafe of the venom, and
their legs became livid, fwelled, and hard.

*What is the Quantity of Venom required to kill an
Animal ?*

We may, I think, from the above experiments,
determine with fome probability, the quantity of
venom it requires to kill an animal.   This queftion
already begins to become important to us on our
own accounts, fince we may at length be enabled to
flatter ourfelves, that the bite of the viper is not fo
dangerous as we have been hitherto taught to imagine it to be.

We have juft feen that the $\frac{1}{1000}$ part of a grain
of venom, introduced immediately into the mufcle
by an incifion, may be a fufficient quantity to kill a
fparrow, although this animal does not always die
in confequence of the introduction of fuch a quantity ; and that it requires about four times as much
to kill a pigeon.   We may even fuppofe it to require about fix times as much to kill the laft animal to a certainty.

The

The fparrows on which I made my experiments weighed fomewhat lefs than an ounce each, and the pigeons fomewhat more than fix ounces each. Now let us fuppofe that fparrows weigh exactly an ounce, and pigeons exactly fix. The quantity of venom it will require to kill a large animal, an ox for inftance, fuppofing it to weigh 750lb. will be about twelve grains ; and it will require nearly two grains and a half to kill a man, fuppofing him to weigh the fifth part of what an ox weighs, that is to fay 150 lb.

It is true that this calculation takes for granted fome new hypothefes more or lefs probable, but of which neither is unlikely. A fufficient number of experiments are wanted, to render them either ab-folute truths, or fufceptible of fome reftrictions.

The firft hypothefis fuppofed here is, that the ve-nom of the viper acts on an animal in proportion to its quantity. There is nothing unreafonable in be-lieving this to be the cafe, fince if a very fmall por-tion of venom is capable of deranging the economy of an animal *to a certain point,* a greater dofe of it ought to produce a greater derangement, a more violent difeafe. Befides we have feen, that animals bit feveral times by one viper, or by feveral, die fooner than thofe that are only bit once by a fingle viper; and we know that a viper which bites feve-ral times, introduces frefh venom into the part at each bite.

The fecond hypothefis is, that the diforder pro-duced in the animal economy by the venom of the
viper,

viper, is lefs in proportion, or rather that the power of the animal to refift the action of the venom, is greater in proportion, as the animal is larger. This is generally fo, although there may be exceptions to this law, that may prevent its being rigoroufly the cafe.

The third hypothefis is, that from the effects produced in an animal of one fpecies, we may argue as to the effects produced in an animal of another fpecies; that is to fay, from birds to quadrupeds. This argument is drawn from a fimple analogy; but this analogy is at the fame time formed betwixt animals with warm blood, and it may therefore be deemed of fome weight.

Now granting that a viper of a middle fize has in its veficles two grains in weight of venom, it will require the venom of fix vipers to kill an ox, and nearly of two to kill a man.

But if we reflect that a viper which bites, does not leave itfelf without venom; that at each bite, at leaft for the firft three or four, it may bring about the death of an animal with almoft the fame facility; it will not appear altogether unlikely, that it may perhaps require twenty vipers, each biting only once, to kill an ox, and five or fix, with the fame reftriction, to kill a man.

CHAPTER

CHAPTER II.

*Of the Time it requires for the Effects of the Venom of
the Viper to become sensible.*

A QUANTITY of the venom of the viper which
scarcely weighs the $\frac{1}{300}$ part of a grain, produces,
on being introduced into the body of a small animal, so violent a disease, that death follows in a few
minutes. It is therefore very clear, that it must
possess a great degree of activity, and that its effects
must be both sudden and powerful. I have advanced in several parts of this work, that the venom of the viper renders the parts that have been
bit, in animals, and that almost in an instant, incapable of exercising their usual functions. I am at
least certain, that I have observed this effect in several that I have had bit. It has been seen that the
venomed part becomes livid after it has received
the bite, but that this does not happen till within
a certain space. The wounded parts soon become
swelled and painful, and the adipose membrane is
shortly after filled with a black and dissolved humour; whilst the blood that remains in the vessels
is black and coagulated.

One would naturally suppose, that the action of
this venom on the organs of an animal is momenta-

3   neous,

heous, and that it is not different from that which takes place when two fubftances of different natures are blended together, and of which chemiftry furnifhes us a thoufand examples.

Defirous of purfuing this idea, and flattered with the hope of difcovering fome effect, or fome particular, that might be ferviceable to my prefent refearches, I formed a new plan of experiments.

My firft trials had for their principal aim, the obferving of the changes the venom of the viper would produce, when introduced into a part cut from an animal, but ftill warm and palpitating.

Experiments on the Limbs of an Animal, recently feparated from the Body.

At the very inftant the part was cut off, I had it bit by a viper, fo that when the experiment fucceeded well, as it frequently did, there could fcarcely pafs a fecond betwixt the amputation, and the bite.

I made choice of young pigeons for this experiment, becaufe I had obferved in thefe animals, that the venom of the viper very fpeedily produces a livid fpot, in the part of the mufcles through which it has introduced itfelf.

To make this experiment, a perfon holds the animal in one hand, and in the other a pair of open fciffars, betwixt which is the leg of the pigeon to be cut off. Another perfon holds this leg in one hand,

U   and

and in the other the head of a viper with the teeth laid bare, and forces thefe teeth deeply into the mufcles of the leg, the inftant it is feparated from the body. The head of the viper has been feparated from the body fome minutes before, and, to make the experiment more commodionfly, has been deprived of the lower jaw. This head is ftill alive, and the fmalleft compreffion that can be made, is fufficient to make it of itfelf draw its teeth from out of their bag or fheath, and force them into the parts that are made to approach it.

It is certain that there never paffed, in any one of the dozen experiments I firft made, more than three feconds betwixt the amputation and the bite; feveral of thefe experiments were made in a fingle fecond, or precifely at the very inftant of cutting off the limb.

In fome of the legs that were cut off, the venom was feen furrounding the holes made by the teeth; in others it was feen flowing out of the holes; and in others no venom was to be perceived exteriourly. On examining the mufcles bit in this way by the viper, I could difcover no fign of a communicated difeafe, neither could I obferve any fupervening lividity about the holes. The blood continued in a fluid ftate in the veins and arteries.

Thefe legs, which were yet warm and palpitating, and which bled, being kept for minutes, and even for hours, afforded me nothing further that was worthy of obfervation.

I re-

I repeated this experiment on the bared, and almoft pale and tranfparent, mufcles of twelve frogs.
The event was exactly the fame ; there was not the
fmalleft apparent fymptom of communicated difeafe.

I repeated thefe experiments afrefh, as well on
pigeons as frogs, having the amputated legs bit by
frefh vipers, previoufly well irritated. The refult
was the fame in all.

I got ready the legs of pigeons and frogs, and as
foon as they were cut off, wounded them with teeth
taken from the head of a dried viper. The fymptoms that refulted from thefe fimple mechanical
wounds, were not fenfibly different from thofe of
the wounds into which the venom had been introduced, although made within the fame time.

It feems then to be an eftablifhed truth, that the
venom of the viper produces no fenfible change in
parts feparated from an animal, notwithftanding they
are yet in a ftate of palpitation. This truth appears
to me of the higheft importance in eftablifhing the
theory of the venom, and deferving of the utmoft
attention.

In the firft place, it is certain, as I have particularly affured myfelf, that there ftill fubfifts in the
amputated leg, for upwards of twenty feconds,
the degree of heat it had before it was cut off.
A perfect irritability is ftill retained in the mufcles,
which continue to move, even for whole minutes.

U 2   The

The arterial and venous fluids ftill remain in tne part, at leaft in a great meafure, and they ftill keep in motion there during fome time.

Thofe who have examined the circulation of the blood in cold adimals, know that this fluid ftill continues to circulate for a long time, in the parts of thefe animals that have been cut off.

Notwithftanding this, the venom feems to be entirely inactive and innocent, in all the cafes I have related above, although every thing fubfifts in the part bitten ; that is to fay, humours, arteries, veins, nerves, irritability, and motion.

This circumftance appeared to me fo new, and at the fame time fo paradoxical, that I was defirous of trying a new kind of experiments, in which the amputated part of the animal fhould approach ftill nearer to its natural ftate, at the moment of being bit by the viper.

I divided the mufcles, nerves, and blood-veffels, of a pigeon's leg with a fharp knife, and left the bone untouched. The incifion was made at the beginning of the *tibia*, immediately below the *femur*. At the fame inftant, I had the flefh beneath the incifion bit by a viper.

But in fpite of all this precaution, I could not perceive, either that the mufcles were rendered livid, or that the difeafe of the venom had been communicated to them.

I repeated this experiment on eleven other pigeons, and, although I kept them alive a great
while

while, could never obferve any fymptom which in-
dicated in the fmalleft degree, that they would have
died in confequence of the operation.

We may therefore regard this circumftance,
however paradoxical it may appear, as beyond all
kind of doubt. On obferving it, I began to flatter
myfelf that fome truth in phyficks, relating to the
mechanifm of the venom of the viper, might be
drawn from it ; and that we might likewife gather
from it fome principle that would be ufeful to the
comprehending of animal motions. In the firft
place it is certain, that the venom, as far as can be
obferved, does not act by a fimple mechanical motion, or by a fimple mixture of fluids ; fince, if that
were the cafe, as the mufcles were provided with
both the accuftomed humours and motions, it ought
to have produced its ordinary effects in the inftances
related above. Neither does its action feem to depend on an effect in chemiftry, fuch as is brought
about, for example, by the contact of an acid with
an alkali ; and precifely for this reafon, that no effect is produced, although the venom is in contact
with the humours of the leg of the animal.

Experiments

Experiments to determine the Time the Venom of the Viper requires to produce its Effects, after it is introduced into a Wound.

To have excluded an hypothefis of any kind on the mode of action of the venom of the viper, may without doubt be a ftep towards the truth ; this is not, however, fufficient to inftruct us how, and on what part of an animal, it acts. My curiofity was therefore rather excited than fatisfied, and I immediately began to confider how I ought to purfue my experiments,

I reflected that, if the venom of the viper produces no effect on a detached part of an animal, however near it may be to its natural ftate, it is certain that it produces very violent, and very fudden effects, on parts that have not been yet feparated.

The firft enquiry that naturally prefented itfelf, was to know whether this venom produces its ufual effects, or rather, whether it communicates its difeafe to the part bit, at the inftant, or not till the end of a certain time.

With this view, I enraged a large viper, and made it bite the leg of a pigeon twice, the fecond bite inftantly fucceeding the firft. I immediately cut off the leg, and examined it with attention, It was very eafy to diftinguifh in it the holes made by the teeth ; but although I kept it a great while, I could never difcover any mark of difeafe or lividnefs.

I had

I had fix other pigeons bit in the fame way, each repeatedly by a fingle viper, and almoft immediately after cut off the leg that had been bit. There was very little difference in the time of my doing this, to all the fix. As there appeared no fymptom of difeafe in the part, it follows as an inconteftible truth, that the venom of the viper does not act inftantly on the part that has received the bite, but that it requires a certain time for this purpofe; fince it is well known, that the parts wounded by this animal, ultimately become livid and fwelled.

The fpace of time it requires to act, was to be determined by experiment.

For this purpofe I had a dozen pigeons bit, each once by a diftinct viper. I meafured with a watch the feconds that paffed betwixt the bite of the viper, and the fucceeding amputation, and managed in fuch a way, that the intervals of time increafed in a ratio of ten feconds; fo that the legs were cut off at the end of 10, 20, 30, 40, 50, 60, 70, 80, 90, 100, 110, 120, feconds after they had been bit. I had previoufly ftripped the fkin from the mufcles, without cutting or lacerating them; and wiped away the blood that flowed from them after the incifion, with a wet fponge. In the leg of ten feconds, I could perceive no change, nor any livid fpot; but in that of twenty there were fymptoms of difeafe. I conceived, at leaft, that I faw an incipient lividnefs about the holes made by the teeth of the viper. In all the others, the difeafe of the

U 4

venom

venom was fo decidedly apparent, that not the fmalleft doubt could remain on the occafion.

I repeated this experiment on twelve other pigeons; but inftead of making the intervals of time betwixt the cutting off of the legs in an increafed ratio of ten feconds, I made them in an increafed ratio of feven.

The leg cut off after feven feconds had no appearance of difeafe. That of fourteen was in the fame found ftate; but all the others, beginning at that of twenty-one, had marks of lividnefs about them. The livid fpots were in general greater in proportion to the delay that was obferved in the amputation. This rule was, however, not without fome exceptions, occafioned by a great variety in the circumftances, which, as any one may readily conceive, are never exactly the fame.

To obtain a more precife information of the time in which the difeafe is communicated, I had twelve other pigeons bit, and in cutting off their legs obferved a ratio of from five to fix feconds, beginning with five.

There was fome doubt in that of twenty feconds, but the difeafe was certain in that of twenty-five. Thofe of five, ten, and fifteen, were without any marks of difeafe, or the fmalleft livid appearance.

A certain conclufion may, I think, be drawn from thefe repeated experiments, that the action of the venom of the viper on the part bitten is not inftantaneous, but that it requires a certain time for its effects to become fenfible in that part.

4

The

The fpace of time that elapfes before the venom
gives manifeft tokens of the difeafe it produces, is
from fifteen to twenty feconds, or thereabouts,

We muft naturally conceive that this time varies
in different animals, and that the difeafe difcovers
itfelf fooner in fome, and later in others.   The peculiar conftitution of the animal, and its fize, ought
to make a fenfible variation, and to modify in a
greater or lefs degree, the action of the venom of the
viper.

But it is fufficient for us to know, that this venom does not operate inftantaneoufly, and to be in
fome meafure acquainted with the time it requires
in acting on certain fpecies of animals.   Thefe *data*
open the way to further refearches.

Is it by the fimple local Difeafe, or by a Diforder excited
   in fome of the moft effential Principles of Life, that the
   Death of the Animals bit by the Viper is occafioned?

The firft enquiry that prefents itfelf, and which
is very important, is to know whether the venom of
the viper produces a difeafe, independant of that
which difcovers itfelf in the part of the animal that
has been bit ; that is to fay, whether it deranges the
animal economy in fuch a way, after a bite has been
received in any particular part, that the animal may
die in confequence of fuch a derangement alone.

I have feen animals, even pretty large ones, fuch
as dogs, fall proftrate on being bit by a viper, without

out being able to ftir for fome time, and with a
fcarcely fenfible refpiration. I have feen others
void their urine and excrements at the very inftant,
as if their fphincters had become paralytick at the
moment of their being bit. It is not a rare cafe to
obferve men fall into a fwoon almoft immediately
after they have received a bite from a viper. But
the agitation of certain animals, and the fear of
others, may contribute a good deal to the producing of thefe effects; and fince it is invariably the
cafe that there is ftill a communication of organs,
and a continuation of humours, betwixt the animal
and the part that has been bit; we may miftake for a
communication of difeafe, what is no more than a
fimple correfpondence betwixt the part bit, and the
other parts of the animal. After all, it muft be left
to experiment to decide on this, as well as on every
other point.

I had a pigeon's leg bit repeatedly by a viper,
and cut off the part foon after at one blow, at the ar-
ticulation of the femur with the tibia.

The leg, when cut off, had all the fymptoms of
difeafe; the holes made by the teeth of the viper
were livid, and the ufual fmall fpots were diftinguifhed. The pigeon died at the end of four minutes.

I had remarked, in making the experiments related above, that the amputation of the leg is not
mortal to pigeons; at leaft, I found feveral that
were deprived of that part, ftill living at the end of
feveral hours.

To

To prevent the following experiments from being in the leaft equivocal, I cut off in the firft place the legs of fix pigeons, a leg from each, to ferve by way of comparifon to the others.

I had twelve pigeons bit fucceffively, fome once, others feveral times. Betwixt the bite and amputation, there did not elapfe in any one of them, lefs than one minute, and more than two. All the pigeons died, and the times of their death are expreffed by the following numbers, denoting fo many minutes, 2, 2, 3, 4, 4, 4, 7, 7, 10, 12, 12, 14.

Of the fix pigeons mentioned above, the legs of which I had cut off without having them bit, neither died, nor did either of them appear to have fuffered in any fenfible degree. I let them live eight days, during which time they fed as ufual, and they then ferved me for other purpofes.

Thefe firft experiments fhow, and that in an unqueftionable way, that a mortal difeafe is communicated to the animal in a very little time; and that it dies, independently of the local difeafe, by an interiour derangement, which the venom has already communicated to its whole fyftem.

This new difcovery was of too much importance not to require ftill further experiments.

I had twenty-four pigeons bit by as many vipers, and at the end of a minute, or with very little variation, if any, from that time, cut from each of them the leg that had been bit. They all died, at the times expreffed by the following numbers, denoting

ing fo many minutes, 3, 3, 3, 4, 4, 5, 5, 7, 7, 7, 7, 9, 9, 10, 10, 10, 10, 10, 12, 12, 13, 13, 14, 20.

It is certain, as I have fince affured myfelf by fresh experiments, that the amputation of the leg is not only not mortal to pigeons, but that it does not feem to be productive in them of any kind of complaint. It is equally certain, as we fee by the experiments related above, that the pigeons bit in the leg by the viper die, notwithftanding the part is removed, provided the amputation is delayed till the end of a certain time. It is therefore a demon- ftrated truth, that the venom of the viper excites in an animal that has been bit, a difeafe independent of the part bitten; and that the animal dies of this fecond difeafe, and not of the local difeafe of the leg; fince the latter fubfifts no longer when the part is cut off, which does not however prevent the death of the ani- mal. This at leaft has unqueftionably been the cafe in the pigeons on which the above experiments have been made. But what is ftill more extraordinary, is that thefe animals die much fooner when the ve- nomed leg has been removed, than when it has not. We have already feen, that in pigeons the fimple amputation of the leg is of no confequence, and it is therefore very furprizing that the local difeafe, which becomes extremely violent, being removed, this circumftance, inftead of retarding the death of the animal, rather accelerates it. This would lead one to fupect that the part bitten ferves to divert the vitiated humours in the animal, and that it is, if I may fo exprefs myfelf, a difeafe excited by the

                                                    animal

animal itſelf, or rather by that principle which exiſts in a living animal, and which, agreeably to the opinions of Hippocrates and Sydenham, ſeems to preſide over its life, and to be the moderator of it.

Is the internal Derangement which the Venom of the Viper cauſes in Animals that are bit, produced at the Inſtant of the Bite, or ſome Time after ?

What is now of the greateſt conſequence to be known, is whether the diſeaſe of the venom is communicated to the animal inſtantly, or not, on the introduction of the venom itſelf.

We have already ſeen what the local malady is, and what are the ſymptoms of it; the time has likewiſe been determined that the venom requires to produce any ſenſible effect on the part bitten. The internal diſeaſe is that which becomes univerſal in the animal, and which even occaſion its death, independently of the external and local diſeaſe juſt mentioned.

To determine whether the internal diſeaſe is inſtantaneous, or not, I made the following experiments.

I had a dozen pigeons bit in the leg by as many vipers, and cut off the part immediately after, in each of them, at a ſingle blow. There was not more than three or four ſeconds betwixt the bite

and

and the amputation. Neither of the pigeons died, nor had any fymptom of difeafe.

I repeated this experiment on twelve other pigeons, which were likewife bit and mutilated within the fpace of three or four feconds. Neither of them died, nor had the leaft apparent illnefs.

It is therefore certain that the venom of the viper does not produce the internal difeafe inftantaneoufly, but that it requires a certain time to communicate itfelf to the animal. We are now to enquire what that time is. Is it the fame as that which it requires to produce the external difeafe?—If this is the cafe, by what common principle do thefe two effects go hand in hand together? And why may not the external difeafe be anteriour to the internal one? The venom begins by touching the local part, and previoufly mixes with the humours of that part.

But let us proceed to experiment.——I had a dozen pigeons bit, each once in the leg by a diftinct viper, and cut the leg from each, with an interval of five feconds betwixt the refpective amputations. The firft leg was taken off at the end of five feconds. The others at the times expreffed in feconds by the following numbers, 10, 15, 20, 25, 30, 35, 40, 45, 50, 55, 60.

That of fixty feconds died at the end of feven minutes; that of fifty-five at the end of fix; that of fifty at the end of feven; that of forty-five at the end of fix; that of forty at the end of twenty; that of thirty-five at the end of an hour; that of thirty at the end of three hours; and that
of

of twenty-five at the end of ten hours. Thofe of twenty, fifteen, ten, and five feconds, neither died, nor feemed to fuffer in any fenfible degree.

However irregular the time of death in thefe animals may appear, we, notwithftanding, remark, in one particular fenfe, a degree of regularity. Neither of the pigeons died on which the amputation had been made before twenty-five feconds; and neither of thofe recovered, the legs of which had been cut off, on or after twenty-five feconds.

We likewife obferve, that in general the pigeons on which the amputation was made the lateft, were thofe that the fooneft fell victims to the difeafe of the venom.

I was defirous of repeating this experiment of twelve other pigeons, obferving the fame intervals of time. The confequences were it is true fomewhat different; but there was ftill a great regularity betwixt the time of the amputations, and that of the deaths.

The pigeons on which the operation was made at the end of 5, 10, 15, feconds, recovered. That of twenty died at the end of feven minutes, and that of twenty-five furvived. Thofe of thirty, thirty-five, forty, forty-five, fifty, fifty-five, and fixty, all died; and the times of their death, beginning with that of fixty, and going back, are 5, 10, 7, 7, 6, 40, minutes, and eight hours.

Here again we obferve, that neither of the pigeons the leg of which was amputated before twenty feconds, died; and that only one of thofe

lived

lived in which the operation was performed at twenty feconds, or afterwards. They in general died the fooner, in proportion to the delay obferved in the cutting off of the leg.

The pigeon that died, notwithftanding it was mutilated as foon as twenty feconds, made me fuf-pect (as in the former cafes, neither of thofe that had been fubmitted to the operation at this period died) that the fize of the viper, and ftill more the circumftance of its having been irritated, might, partly, however, have occafioned this difference.

To be certain of this, I had two pigeons, perfect-ly alike in fize, bit, one of them by a large well-en-raged viper, the other by a fmall one, that was not irritated. I cut off a leg from each pigeon at the end of twenty feconds. The firft died at the end of five minutes ; the fecond had not the fmalleft fymptom of difeafe.

This experiment convinced me, that the time in which the internal difeafe is communicated, may be greater or lefs, according to the different circum-ftances, that the vipers, and the pigeons or other animals, may be in at the time, and according to the manner of biting.

To affure myfelf ftill more fully of this circum-ftance, I had two other pigeons bit, one by a very large viper, the other by a very fmall one. The firft was enraged, and hifled at the time of biting. The other was made to bite, without being pro-voked in the leaft. The amputation of the leg was made in both of them at the end of fifteen feconds.

feconds. The firft pigeon died at the end of nine minutes ; the other had not the fmalleft complaint.

It follows from all that has been obferved, that it requires a certain time for the venom of the viper to be communicated to an animal, and that this time is fomewhere betwixt fifteen and twenty feconds.

It has been feen above, that it requires pretty much the fame time for the external difeafe to be communicated to the part bitten ; and hence it appears, that thefe two difeafes accompany each other, and that the venom produces both within the fame fpace of time.

This agreement of difeafes and effects, which has thus far appeared fo very regular and conftant, fully deferved to be confirmed by a continuation of experiments, ftill more precife and fimple than the preceding ones.

Of the Symptoms which characterize the Difeafe.

The difficulty confifts in determining the death or difeafe of the animal by the fymptoms that appear in the part bitten ; and *vice verfa*, in determining the fymptoms of the part bitten by the death of the animal. On one hand, thefe fymptoms, as has been remarked before, are neither equivocal nor difficult to obferve ; and on the other, the death of the animal, in confequence of the introduction of the venom, is a truth eftablifhed by experiment.

It would be long and tirefome to enter into a detail here, of the diftinct confequences of the experiments, more than eighty in number, that I made with this view. It will be fufficient for me to fay in general, that neither of the animals on which they were made (except one indeed, the cafe of which was doubtful) died without manifeft fymptoms of the difeafe of the venom in the part bitten; and that (except in five inftances only) I obferved in all the others, that when the animal recovered, there was no fymptom of the local difeafe of the venom. The few exceptions that occurred, which might have depended on a thoufand accidental caufes, do not render the law that thefe two difeafes obferve, nor the conftancy with which they are at the fame precife point of time excited in the animal, lefs certain.

This agreement, fo conftantly obferved, made me fufpect ftill more the exiftence of a certain principle in the animal machine, which prefides and watches over life.

Scarcely has an animal encountered any thing that troubles and deranges the functions of its life, than a new force feems at the fame time to be excited, and to be, as it were, awakened, which endeavours ftrenuoufly to keep the caufe of death from the organs that are the moft effential to life, and to carry the morbifick matter to the part that is the moft difpofed to receive it, whether on account

•ount of wounds that have previoufly been made in it, or of humours that are extravafated by the rupture and laceration of veffels.

The venom of the viper occupies but a very fmall fpace in the leg·of an animal, and may, if one wifhes it, be driven into fo narrow a compafs, as fcarcely to occupy the hundredth part of a line in fuperficies, without any phyfical or fenfible folidity.

Now granting the fuppofition that this fmall quantity of venom is entirely abforbed, and carried into the torrent of the circulation, it ought to be equally diftributed in the mafs of humours of the animal, to the fize of which, or to its veffels, the diftribution of it ought to be proportioned.

But it is quite the contrary; the humours and the blood are carried tumultuoufly, and in hafte, to the part that has been bit, and the blood not only collects about the fimple mechanical wound made by the tooth, but fpreads to a great diftance, and changing its colour, pours in torrents into the adipofe membrane, whilft another part of this fluid penetrates in a diffolved ftate through the coats of the veffels.

It therefore appears, that all the efforts made by an animal which has been bit by a viper, are directed to the difcharging of the blood and humours that are affected by the obnoxious principle the latter conveys by its venom, and to the throwing of them, as much as it can, on the part that has been bit.If it fucceeds in t his way, in fupporting

the

the highly neceffary functions in the vital parts, it furmounts the very fudden and dangerous internal difeafe that would otherwife have been deftructive to it.

As to the external difeafe, the circumftances are altogether different. It becomes fimilar to many other difeafes caufed by an obftruction of humours in the veffels, of fluids extravafated in the adipofe membrane, and of blood which threatens gangrene and fphacelus. If the animal is very ftrong, however great the local difeafe may be, it at length recovers; and I have obferved monftrous fwellings, enormous extravafations, and an entire lividnefs and gangrenoufnefs of the parts, and, notwithftanding all this, the animal has got about again. This is frequently obferved in the larger fpecies' of animals, fuch as refift for feveral days the action of the venom.

I wounded the crural mufcles of three pigeons with venomous teeth, and, almoft at the very inftant, cut off the leg in each of them. The mufcles of the firft pigeon's leg had no apparent fymptom of difeafe. Thofe of that of the fecond had a fmall red fpot, which penetrated through the fibres without changing its colour. In thofe of the third pigeon's there was a fmall red fpot, fimilar to the former one, which penetrated to the tibia itfelf, where it appeared fomewhat darker than ufual.

I wounded the crural mufcles of two other pigeons, with teeth which had been dried a long time, and which I had previoufly well wafhed, and
a moment

a moment after cut off the leg in each of them that had been thus punctured. In one of thefe legs there was no fymptom of difeafe or wound ; in the other there were two red fpots, which penetrated into the mufcles, infenfibly lofing their rednefs.

I wounded the crural mufcles of three other pigeons with venomous teeth, and bound and cut off the legs at the very inftant. In one of thefe legs there was an appearance of black and extravafated blood. The fymptoms of difeafe in the other two were perfectly vifible and certain ; that is to fay, a livid colour, and black and extravafated blood for the whole depth of the mufcle.

I wounded the crural mufcles of two other pigeons with dried teeth, and at the fame time bound and cut off the legs. The blood was extravafated in both, and was become of a dark colour,

Experiments to determine whether at the Moment of Amputation, a fubtile Principle of fome Kind does not efcape from the Blood.

The little conftancy I met with in thefe experiments, and the fufpicion that a volatile fluid of fome kind might have efcaped from the blood on the moment of its being difcharged from the veffels, and expofed to the open air, induced me to engage in fome other trials, which I conducted in the following manner :—I held the pigeons in fuch a way, that although their legs were perfectly dry, their

thighs

thighs were entirely plunged in water. The amputation was made beneath the water, in the thigh, ſo that the inciſed part could have no communication with the air ; and the muſcles were wounded under water with venomous teeth. This being done, I kept the foot under water for three or four minutes, drew it out again, and examined it.

I repeated this experiment on the ſame number of pigeons, and ſimply wounded their muſcles with dried teeth. There were marks of the ſimple mechanical wounds, not only in the venomed muſcles of the pigeons in the former experiment, but likewiſe in thoſe that had not been venomed of this one. As I found no difference betwixt them, I cannot take upon me, with any degree of probability, to eſtabliſh a fact of any importance on theſe appearances.

I examined ſeveral times into the parts adjacent to the one that had been bit, either in animals which were already recovered, or in thoſe in which there were no longer any certain ſymptoms of diſeaſe, and of which the parts had almoſt regained their uſual motions. I obſerved with ſurpriſe in ſeveral of theſe animals that had been bit in the leg, that there was ſtill a great extravaſation of humours in the adipoſe membrane, at a very great diſtance from the part that had been bit ; and likewiſe that all the abdominal muſcles were ſtill red and inflamed. Every thing in ſhort concurred to perſuade me of the exiſtence of that principle, which

which has been either fufpected or admitted by others ; and to convince me that the local difeafe is not a mechanical effect of the introduction of the venom into the part, but is rather the means this vital principle employs to drive towards the exte- riour parts the morbifick matter that circulates in the humours, and to relieve from it the organs that are the moft effential to the prefervation of the animal. I fhall point out at the conclufion of this work, the purpofe that may be drawn, and the uti- lity that may be derived, from this diftinction of the two difeafes that the viper occafions in an ani- mal by its bite. The want of attention to thefe two ftates of the animal, fo different from each other, has thrown the greateft perplexity on this fubject, and has enveloped it in errour and obfcu- rity. That which belonged to the one has been afcribed to the other, and thus has every thing been confounded.

C H A P T E R   III.

On the Action of the Venom of the Viper upon the Blood
of Animals.

IF the matter of the preceding chapter has been of
some importance, which cannot be denied; if it has
presented new and altogether unexpected phenomena; if it has been a guide to us in establishing
principles and vital powers in the living machine;
the subject of the following chapters will certainly
not be less important, whether we regard the novelty of their contents, or the use and applications
that may be made of them, in obtaining a knowledge of the venoms that are analogous to that of
the viper, and in explaining the animal mechanism,
as well in a state of disease as in perfect health.

Mead, to determine whether the venom of the
viper had any degree of action on the blood of an
animal that had been bit, mixed five or six drops
of it with half an ounce of blood, in the colour
and consistence of which he could observe no
change, as the consequence. There was in short
no difference betwixt this blood and another quantity drawn at the same time, which he had put into
a vessel similar to that which contained the first, by
way of comparing the two together. This experiment

ment I repeated, and received the blood which flowed from the divided veſſels of an animal, immediately into a concave glaſs, which I had previouſly warmed, and into which I had put five grains in weight of the venom of the viper. The paſſage of the blood from the veſſels to the glaſs was ſo quick, that it is not poſſible to have it out of the veſſels, in a condition approaching nearer to its natural one. On the moment of the union of the venom with the blood, I obſerved the latter, the quantity of which was about an ounce, or ſomewhat more, with a very ſtrong lens. I could never perceive any kind of motion in it, neither could I diſtinguiſh in it any diſſolution, nor the ſmalleſt appearance of coagulum ; in a word, it was entirely in its natural ſtate. Its globules were of their uſual ſhape, and its colour was equally preſerved. This particular ought not to ſurprize us, after the experiments that have been made on the legs of pigeons bit by the viper at the very juncture of their being cut off; and likewiſe after thoſe in which they were cut off ſometime after they had been bit. The blood in theſe caſes certainly approaches much nearer to its natural ſtate than when it is drawn from the veſſels. There is here both the natural heat and ordinary motion of the humours, and in ſhort the life of the organs themſelves.

Nothing appears more natural than the deducing from theſe particulars, that the venom of the viper has no action on the blood of the animal that receives the bite. This is indeed the inference that

Mead

Mead has drawn from the above recited experiments on the blood of animals taken warm from the veffels.

However perfuafive this experiment on the blood might have been, and however refpectable the authority of Mead, I determined to try a new kind of experiments, partly analogous to thofe related above, but more direct and more fimple. Thefe experiments confift·in introducing the venom of the viper in an immediate way, without touching any of the parts that are previoufly cut, into the blood. They are indeed fomewhat difficult, but are ftill poffible, and are made by injecting the venom of the viper, by means of a fmall glafs fyringe, into a vein that has been opened with a lancet. I forefee an objection that will be made, that experiments of this kind are altogether ufelefs after thofe that have been related, to which they are befides perfectly analogous; and that feeing there has been no change obferved in the venomed blood, it ought, from a parity of reafoning, to be concluded, that there will be no greater change in it, in thefe experiments. Such is the rifk of being miftaken, that thofe incur who love rather to reafon than to experiment; and this is the mode of arguing of thofe philofophers, who, perfuaded that they are arrived at the fountain-head of natural fciences, flatter themfelves that they know every thing, and are capable of explaining every thing.

Injection

Injection of Venom into the Blood Veſſels; and its Effects.

The experiments I am about to relate, were made on large rabbits. The jugular vein was the veſſel on which I operated.

When a great portion of hair has been removed from the inferiour part of the ſide of a rabbit's neck, and a large inciſion made in the ſkin, the jugular vein is diſcovered dividing itſelf into two ſmaller branches. I ſtrip in the experiment, the two branches and a part of the trunk of the jugular vein, of about ten or twelve lines at leaſt in length, of the adipoſe membrane, and the other neighbouring parts. I tie one of theſe branches with a thread, at the diſtance of ten lines from the trunk, and tie another thread to the ſame branch, about ſeven lines below the firſt, ſo that this ſecond thread is only three lines from the trunk. This laſt thread has a knot, ready to be drawn tight at a proper time. But before I go any further, I think it neceſſary to explain the manner of making uſe of the ſmall ſyringe, intended to convey the venom into the veſſels.

This is a ſmall common glaſs ſyringe, terminating in a capillary tube of ten lines in length, and crooked. I put into this ſyringe the venom that I mean to introduce into the vein. I uſually cut off two vipers' heads, and receive all the venom

from

from their veficles in a fmall cryftal fpoon. I add
to this venom the fame quantity of water, and when
the liquors are well blended together, draw them
up by fuction into the fyringe. There ufually enters into the fyringe at the fame time, a fmall air-
bubble, which is eafily difperfed, by pufhing the
pifton forward a little towards the tube. The fmall
quantity of the liquor that flows with the air out of
the point of the tube, is received in the fmall fpoon,
and is fucked up again into the fyringe, by once
more withdrawing the pifton a little.

The fyringe being thus freed of the external air,
I withdraw the pifton in an almoft infenfible degree.
The venom retreats a little, and leaves the point of
the capillary tube, which remains full of air, for the
length of four lines. The quantity of this air is
very trifling, on account of the fmallnefs of the diameter of the tube in that part. I now wipe the
crooked part or extremity of the fyringe with a piece
of very fine moiftened linen, and introduce a very
fine and dry linen thread, for the length of two
lines, to cleanfe the venom, and likewife the fmall
fpace in the capillary tube, that is occupied by the
air.

The fyringe being thus in readinefs, I raife a little, by the uppermoft thread, the branch of the jugular vein to which the two threads are faftened, betwixt which I open it with a lancet, and introduce
the capillary extremity of the fmall fyringe at the
orifice, continuing this till it has entered four or
five lines into the principal trunk. I now draw the

end

ends of the threads together, the lower one of which binds the coats of the veffel very ftrongly to the capillary tube of the fyringe. Things being in this ftate, I pufh forward the pifton of the fyringe by degrees, and force out of it all the venom, which paffes entirely to the trunk of the jugular vein, to be carried an inftant after to the heart.

This experiment requires two perfons at leaft, and fucceeds ftill better when there are three. If the fyringe has been previoufly got in readinefs, it does not continue at the moft for more than two minutes altogether; and when the parts of the animal are known, and it has been made a few times, is not fubject to any inconvenience.

Before the fyringe is drawn out of the veffel, I have been accuftomed to withdraw the pifton a little, that a fmall quantity of blood may enter the capillary tube, and that none of the venom may remain at the orifice of it. At the moment of my drawing out the fyringe, I again tighten the lower thread, fo that the veffel remains perfectly clofed. I raife with pincers the portion of the jugular vein betwixt the threads.

It was not without reafon that I made choice of a veffel which branches out into two others; neither was it at hazard that I introduced the capillary part of the fyringe into the principal trunk itfelf.

I wifhed that the venom fhould be carried immediately to the heart, and I could not think of a better expedient than that of procuring a very large lateral veffel, where the blood continuing to run in a

3   full

full ftream towards the heart, muft neceffarily carry
with it the venom that it meets with in the trunk.

Thefe experiments are too important not to be related with fome degree of detail. They at leaft require me to defcribe the principal circumftances by
which they were accompanied. I fhall give them
here in the order in which they were made.

I injected into the outer jugular vein of a large
rabbit that weighed feven pounds, the venom of two
viper's heads, got ready in the manner defcribed
above, and with a nice obfervance of all the precautions I have juft laid down. The venom fcarcely began to enter the vein, when the animal gave feveral horrible cries, difengaged itfelf, writhed itfelf
about, and died a moment after.

The novelty of this ftrange and unexpected event,
prevented me from calculating the exact time the
animal lived after the injection of the venom; neither could I afcertain the time I employed in pro-
pelling the whole of the venom from the fyringe.
It is certain, however, that the animal did not live
more than two minutes, and that the injection was
made within the fpace of eight or ten feconds.

As I was defirous of feeing whether this experiment was a certain one, or whether the animal died
in confequence of fome circumftance I was ignorant of, I examined the ftate of the vifcera in the
dead animal, and likewife that of the blood in its
veffels. I was likewife induced to vary fome of the
circumftances in making the fucceeding ones.

I got

I got ready another rabbit in the above manner, and began by injecting a quantity of water into a branch of the jugular vein, equal to that of the mixture of venom and water in the preceding experiment. The rabbit did not suffer in the least. I kept it in this state during five or six minutes, and perceiving that it did not become at all disordered, sat about injecting into the same vein, the quantity of venom mentioned above.

The animal, however, neither cried out nor was agitated. At the end of a few minutes I perceived that it had sickened, and it died at the end of twelve hours. All the parts I had stripped of the skin, to lay bare the jugular vein, were violently inflamed, and very livid. The adipose membrane was filled with black extravasated blood. All the pectoral muscles at the side on which I had injected the venom, and a part of the abdominal ones, were already become livid. The very intestines were inflamed. The inner part of the thorax was inflamed likewise, and was bloody; and the heart had formed adhesions. The blood, both in the large vessels and heart was coagulated and black ; and the lungs were marked here and there with somewhat livid spots.

This second experiment convinced me of the very great importance of thoroughly examining the state of an animal after its death. It is principally by this state, that we ought to judge of the action of the venom on the blood.

But how came it about that the first rabbit died as it were instantly, and the second not till the end

of

of twelve hours ? To what is this difference to be afcribed ?

I inftantly proceeded to a third experiment, hoping to draw fome further information from it.

I got ready a rabbit, and injected the venom of two vipers as before, into the branch of the jugular vein. The rabbit did not feem to fuffer in the leaft from this operation, and recovered of the external difeafe in a few days, as readily as if it had only undergone the preparation neceffary to the injection of the venom. An hour after this injection had been made, I found it eating, as if in perfect health.

This third experiment completed my perplexity, and I began to miftruft altogether. In the firft place I faw an animal die, as it were at the moment of injection, and diftinguifhed a real difeafe in that which lived twelve hours. It was therefore certain that the venom, when united with the blood, was capable of producing fuch a derangement in the animal machine, as to excite in the animal a very violent difeafe, terminating in death. All this was real; but how could thefe two cafes be reconciled with the third ?

Some doubts occurred to me as to the method I had purfued in making thefe experiments, which had not been altogether conducted with the exactnefs and precifion I defcribed a little time ago. I did not make the fecond ligature in the vein; I did not examine whether the capillary tube reached into the principal trunk; and I did not withdraw the pifton of the fyringe, before I drew the capillary

tube

tube of the latter out of the veffel. The neglect of thefe precautions made me look upon the three experiments I have juft related as fufpicious, and I fat about experimenting afrefh, with a greater de-gree of attention and precifion than before.

For this purpofe I got ready a large rabbit, healthy, and in good plight. I made the two liga-tures in the external branch of the jugular vein, and introduced the capillary tube into the common trunk, tightening the thread on the tube, and in-jecting the whole of the liquor at once. I took care to withdraw the pifton a little, before I drew the fyringe out of the vein, and to tighten the thread once more. In a word, I did not neglect any one of the precautions I had previoufly determined to take. The confequences were as follows.

The whole of the venom had fcarcely paffed from the fyringe into the jugular vein, when the rabbit gave feveral horrid fhrieks, and was feized with very violent convulfions. It died in lefs than a minute and an half. There were not more than feven fe-conds fpent in the injection.

The blood in all the large veffels was black and coagulated. It was likewife fo in the heart and au-ricles. The coronary veins were fwelled and livid; and an extravafated black blood, in large fpots, was feen about them in the mufcular fubftance of the heart. The pericardium was entirely filled with a liquor, fo as to be diftended like a bladder; and this liquor was tranfparent, with a flight red tinge.

Y   The

The lungs were full of the ufual fpots, through which, when they were touched in the flighteft degree, the air rufhed out of the water that covered this vifcus. The inteftines, ftomach, and mefentery, were covered with fmall, livid, and red fpots.

This experiment fucceeded too well to leave me in any kind of doubt as to the nature of its confequences. The animal died almoft inftantly, and cried out the moment the venom had entered the veffel.

The two vifcera that are the principal organs of life, were inftantly attacked by a violent and mortal difeafe. The blood was immediately coagulated in the large veffels, in the lungs, and in the heart. In a word, every thing concurred to the fudden ftoppage of the circulation, and to the death of the animal.

The extravafation of the blood of the coronary veins is furprizing, and the livid fpots of the lungs, and dilacerations of this vifcus ftill more fo. But what fuprized me moft of all, was the blood collecting in fuch an abundance, in fo many veffels, and in fo many cavities. In this difeafe an extreme diffolution of a part of this humour, exuding every where through the veffels, takes place; and, at the fame time, a coagulation of another part, which fixes and condenfes in a few moments.

Every advance I made, in this new career of experiments, prefented me either with fomething paradoxical, or with a novel and unexpected circumftance. I paffed on to the fifth experiment, which
I made

I made exactly as I had done the fourth. Although the result of it was somewhat different from that of the last, it agreed very well as to the nature of the disease, and as to the opinion that ought to be held of the introduction of the venom of the viper into the blood. On the injection being made, the rabbit did not cry out, neither did it seem to suffer in any sensible degree. At the end of an hour, however, it appeared to be sick, refused its food, and died at the end of twenty-four.

On opening its body, I did not find the abdominal viscera to be much inflamed ; but to atone for this, the usual livid spots, and the air gushing freely out of them, were seen on the lungs. All the muscles of the breast were considerably inflamed, and the whole of the adipose membrane, from the neck to the lower part of the belly, was filled with black, extravasated, and fluid blood. The blood in the heart, in the lungs, and in the large venous vessels, was coagulated ; but much less so than in the cases related above in which the rabbit died almost instantaneously.

I immediately proceeded to the sixth experiment, to see whether there would be any degree of uniformity betwixt the injection of the venom, and the death of the animal. I neglected to remark, in relating the preceding experiments, that I had found sometimes a greater, sometimes a less, quantity of venom in the viper's heads, and that in some of them I had even observed a white and somewhat glutinous matter flow from the tooth.

I had

I had likewife found the palate of fome of the vipers I employed inflamed to a certain degree, and the two fheaths or bags of the teeth likewife inflamed and red.

But I could not fay pofitively, whether thefe circumftances could have been capable of influencing the effects of the venom on the animal. I therefore refolved to take it in future from no other heads of vipers, but fuch as were perfectly found, and the beft fupplied ; and to procure it in a greater quantity.

I got ready a large and ftrong rabbit in the ufual way, and introduced into the fyringe the venom of two very large vipers, the heads of which were in a found ftate.

The venom was not yet completely injected, when the rabbit began to fhriek, and died in lefs than two minutes in very violent convulfions. Having opened the breaft, I found the auricles and ventricles filled with grumous blood. That of the large venous veffels was in the fame ftate. There was a great deal of lymph in the pericardium, in which there was likewife extravafated and concreted blood. All the inteftines were in a very inflammatory ftate, as were alfo the ftomach and mefentery. The arteries were in general empty. The lungs were but little fpotted ; but in inflating them beneath the water, the air was feen rufhing out in feveral parts, and the fmall fpots were then apparent. The blood in the lungs was likewife concreted.

I got

I got ready another rabbit, and injected in the usual way into the jugular vein, the customary quantity of venom.

It scarcely began to enter the vein, when the rabbit cried out, and in less than two minutes died, with the most terrible shrieks and convulsions.

I opened it, and found the lungs spotted as usual, and the blood coagulated in the two ventricles. It was much more so in the right ventricle than in the left, as I had also found it in all the preceding cases. It was likewise in the same state in the auricles and veins. The pericardium was filled with water mixed with blood. The coronary veins had, for their whole circumference, two large, longitudinal, and livid spots. The blood in the lungs was black and grumous, and the air gushed out as usual. The intestines were inflamed, as were also all the abdominal muscles; and there was a great deal of extravasated and dissolved blood in the adipose membrane.

These two last cases are very uniform, and agree too well with the preceding ones, to admit of any doubt as to the immediate action of the venom of the viper on the blood.

Further Experiments on the Jugular Vein of Rabbits.

Notwithstanding the uncertainty and obstacles that are met with in experimenting on the blood-vessels, I was desirous of making some further trials

Y 3

on them, conducted with all poffible care and attention, as the very great importance of the fubject feemed to require. I chofe for this purpofe, two of the largeft rabbits I could procure, each of them weighing ten pounds. I took the venom from two found vipers, which I had previoufly examined with great attention for that purpofe. I had not yet finifhed the injection in either of the two rabbits, when they gave feveral loud fhrieks, and died in the moft violent convulfions in lefs than two minutes. Having opened the thorax of each rabbit, I found the lungs fpotted as ufual, and the blood veffels and auricles filled with black and grumous blood. The pericardium, as in the former cafes, contained a humour, and the inteftines and mufcles were, as ufual, inflamed.

The immediate action of the venom of the viper on the circulation of animals with warm blood, is therefore both indubitable and conftant. It is a fact, however, that would not have gained any degree of credit, had it not been for thefe laft experiments, fince it feemed in fome meafure to be contradicted by the former ones, which, although they were lefs direct and lefs fimple, were neverthelefs made on the blood. This fhows us how cautious we ought to be in the inferences we draw from experiments ; and at the fame time proves to us, that we know little or nothing, at leaft with any certainty, and without incurring the rifk of being deceived, beyond that which is demonftrated to us by experiment alone.

But

But in what way are we now to reconcile the immediate action of the venom of the viper on the blood, when it is injected by the veins; and the inactivity of this same venom, not only on the parts of an animal recently cut off, but likewise on those that have remained in an entire state, and united with the animal, during a period of 15 or 20 seconds, after it has been introduced?

I must acknowledge that this is a very great difficulty, and that it would be no easy task to explain it in a satisfactory way. It appears that there can be nothing deficient in the parts that are still connected with the animal when they are bit by the viper. It even seems probable that these cases have an advantage over those in which the trial is made on the blood, since both the muscular fibres and nerves are wounded by the viper's teeth, instead of which the venom, when injected into the vessels, certainly touches neither one nor the other of these parts. What is then the cause that retards the disease of the venom for several seconds in the part of an animal that has been bit; and how it is that the venom does not produce any disease in the parts that are either cut off and bit immediately after, or are cut off immediately after they are bit.

It is probable that there may be an unknown principle in the blood circulating in the vessels, which ceases to exist the moment this fluid is drawn out of them, and which is likewise no longer to be found in the parts that are recently separated from the body of an animal. This principle, granting

Y 4   the

the fuppofition, therefore, poffeffes fuch very active and fubtile qualities, that it is diffipated at the very moment the part is feparated from the animal.

We have feen that the venom fcarcely comes [in contact with the blood in a veffel, when the moft violent derangements are produced. The animal fuffers extremely, and the blood is condenfed in an inftant. If this fame venom is mixed with the blood as it flows warm from a veffel that has been opened; or if it is introduced into any part of a mufcle that has been feparated an inftant before; it produces no effect, and no appearance of difeafe is obferved, nor any condenfation of humours. Here, however, every thing is the fame, unlefs it be that in the cafe in which the venom is introduced into the veffel, there is a blood circulating with the reft of the humours, and always covered by the coats of the veffels; inftead of which, the blood that is drawn from a vein is out of the torrent of the circulation; and that of the parts which have been recently cut off has already been in contact with the air, and the veffels which contain it are open.— However it may be, fince the effects are very different from each other, the circumftances muft be fo likewife; and the only conclufion we can draw, as to the humour contained in a veffel, and the humour drawn out of a veffel, is, that there exifts in the firft cafe, fomething that is not to be met with in the fecond.

Agreeably

Agreeably to this hypothefis, this new principle which exifts or refides in the blood, in the veffels of a living animal, does not produce the fame effects every where equally, and in the fame time. The venom has no fooner united itfelf with the blood in the jugular vein, than the animal is attacked by a very violent difeafe, and the blood is coagulated a very few inftants after. Inftead of which, in the parts that are more diftant from the heart, where the veffels are fmaller, it requires a certain time for the difeafe to difcover itfelf, and for any fenfible change to take place in the part into which the venom has been introduced.

It thererefore appears, that this principle obferves certain laws in governing the animal economy, and that it is itfelf fubject to certain regulations.

In the cafes in which the difeafe is the moft remote from the heart, and the leaft dangerous, the blood coagulates by degrees, is driven back to the parts bitten, and affords time and opportunity for the efforts of nature to overcome the difeafe, and to preferve the circulation in the organs the moft neceffary to life.

But at length what is this new principle, and what are the organs that fecrete it, and convey it to the veins?

In this very difficult enquiry, it appeared to me that experiment could alone furnifh me with fome light, and conduct me to fome new truth. But where are the experiments to begin?

3   CHAPTER

C H A P T E R　IV.

Experiments on the Nerves.

IN the long courfe of my experiments on the venom of the viper, and in collecting together the circumftances and ideas that prefented themfelves, I never loft fight of the principle of fenfation in an animal, which appeared to me to be acted on by the venom of the viper. I have in confequence of this, judged it neceffary to examine the nerves in which it refides, or which are the organ and inftrument of it.

Mead fays, in the introduction to his work on Poifons, that having better confidered the nature and quality of the fymptoms of the bite of the viper in animals, he is certain that this difeafe is altogether nervous, and that it is communicated by the medium of the nerves, and not of the veffels. In confequence of this theory, he has recourfe to the animal fpirits, againft which he believes that the immediate action of the venom of the viper is exercifed. Indeed if we examine the fymptoms that this venom produces in animals, we are eafily led to believe that a difeafe of fuch a nature belongs to the clafs of difeafes which the phyficians ftile

I　　　　　　　　　　nervous.

nervous. In the courfe of my experiments, I have feen a pretty large dog fall down motionlefs, the moment after it had been bit by two vipers. I at firft thought it dead, but at length perceived fome little remains of refpiration, which was, how-ever, fo flight and feeble, that it could fcarcely be diftinguifhed. The dog continued in this lethar-gick ftate for more than half an hour. I have feen feveral others thrown by the venom into very vio-lent convulfions. Vomiting, anxiety, and rage, occur very frequently; the motion of the heart is irregular and convulfive, and the arterial fyftem hard and contracted. In fhort, they die in the midft of the moft unequivocal fymptoms of fpafms and contractions, and, in a word, with the affections that are by the faculty termed *nervous*.

Another idea occurred to me, that perhaps an active principle, a fubtile fluid, is fecreted by the nerves themfelves, which, when it mixes with the blood, contrives in fome way to animate it, to ren-der it vital, and to maintain its fluidity. In this cafe, the action of the venom of the viper may perhaps have been directed againft this principle itfelf; and thus we may explain, why it is that the blood, when drawn from the veffels, and in the open air, is no longer fufceptible to the action of this venom.

Experiments on the Nerves, Spinal Marrow, and Brain, of Frogs.

I opened the belly of a frog, and laid bare the crural nerves. I poured a fmall quantity of venom on thefe nerves, taking care that it did not fpread to the furrounding parts. At the end of two hours I pricked the nerves with the point of a needle, and the mufcles of the foot contracted. At the end of four hours, however, no part of the animal was fenfible to ftimulation. A frog intended for a comparifon, lived twelve hours, notwithftanding I had opened the abdomen, lacerated the inteftines, and punctured the lungs.

I repeated this experiment twice, and the event was each time pretty much the fame. I began, however, on a little confideration, to think the method I had adopted fallacious. It is almoft impoffible to prevent the venom, when applied to the nerves, from communicating to the adjacent parts. In this cafe, the difeafe, or death, of the frog, may be the effect of the communication of venom to the other parts of the animal, and not the confequence of its contact with the nerve itfelf.

I changed my mode of experimenting, but ftill employed the fame animals.

I cut off the heads of two frogs, alike in fize, and touched the fpinal marrow of one of them, but not of the other, repeatedly with the venom. At the
end

end of three hours the venomed frog appeared to be dead, whilft the other continued to live, and to leap about.

I introduced a pin into the fpinal marrow of the frog to which the venom had been applied. Its fore-legs remained motionlefs, but there was a flight tremulus in the feet. The heart and auricles likewife had a fmall degree of motion. In an hour more every part was at reft. The fecond frog leaped about the chamber at the end of twenty-four hours.

I cut off the head of another frog, and introduced a drop of venom into the fpinal marrow. At the end of an hour the frog fcarcely gave any figns of life. On opening the breaft, the heart and auricles feemed ftill to preferve fome degree of motion, which was however perceived with difficulty. A pin introduced into the fpinal marrow, occafioned an almoft imperceptible motion of the fore-legs and feet. The heart, however, on being ftimulated, performed its ofcillations for a long time.

I cut off the head of a frog, and removed a fmall portion of the fpinal marrow. I introduced by the great opening of the vertebræ, a drop of venom. At the end of two hours the frog was to appearance dead. The heart fcarcely preferved fome little re-mains of motion, which were not encreafed by fti-mulation. A pin introduced into the fpinal mar-row was barely capable of exciting a feeble motion in fome of the mufcles.

I cut

I cut off the head of another frog, and having removed a fmall portion of the fpinal marrow, introduced a drop of venom into the great foramen. At the end of three hours the frog appeared to be dead. Having opened the thorax, I remarked that the heart was ftill irritable; a pin, however, that I introduced into the fpinal marrow fcarcely occafioned a fenfible contraction of the feet.

I repeated this experiment on two other frogs, and the refult was pretty much the fame as in the above experiments. The death of the frogs fucceeded the operation and introduction of the venom in a fpace of betwixt two and three hours. The heart was fomewhat irritable, but the mufcles were little fo, or not at all, notwithftanding my ftimulating the fpinal marrow with a needle.

I now thought it proper to make a little variation in my experiments.

I removed a portion of the cranium of a frog, and applied a fmall quantity of venom to the brain. At the end of four hours the frog was dead, and the heart infenfible to every ftimulation. On pricking the fpinal marrow with a needle, not the fmalleft motion was reftored in it.

I opened the cranium of another frog, and applied a drop of venom to the brain. The frog furvived this operation two hours, at the end of which time the heart had ftill retained a flight degree of motion; it was fhrivelled, black, and contracted. On ftimulating the fpinal marrow, there was an almoft infenfible contraction of the mufcles.

I re-

I repeated this experiment on the brain of four other frogs, and the confequences were very analogous to thofe of the two preceding ones. However having removed the cranium of the two frogs without applying the venom to the brain, by way of a comparative experiment, they both died in the fpace of ten hours.

The confequence of thefe experiments appearing neither fufficiently clear nor uniform, I had once more recourfe to the cutting off of the heads, thinking that by dint of multiplying my experiments in that way, I might affure myfelf of the action of the venom on the nerves.

I cut off the heads of two frogs, and applied venom to the fpinal marrow of one of them, but did not venom that of the other. At the end of three hours the venomed frog was to appearance dead; the other was ftill living, and had a free motion in all its parts. I introduced a pin, which had been dipped in venom, into the vertebral opening of the firft frog, and it excited a very feeble motion of the feet, but had no fuch effect on the fore-legs. I fcarcely ftimulated the fpinal marrow of the other frog with a needle, when the frog leaped about brifkly. At the end of the fourth hour there was not the fmalleft perceptible degree of motion in the venomed frog, and neither the heart nor auricles were any longer fenfible to ftimuli. The other frog was ftill leaping about at the end of thirty hours.

I cut

I cut off the head of another frog, and introduced the venom into the fpinal marrow. At the end of two hours the frog was to appearance dead. Having opened the thorax, the heart was motionlefs, and even infenfible to ftimulations. The fpinal marrow, when likewife ftimulated, fcarcely excited any degree of motion in the feet.

I repeated this experiment, obferving the fame circumftances, on another frog. At the end of three hours I found it dead. The heart and mufcles were perfectly motionlefs. On treating another frog in the fame way, the confequences were altogether fimilar.

I cut off the head of another frog, and applied venom to the fpinal marrow. At the end of five hours the frog ftill retained fome feeble figns of life. On opening the thorax, I found the heart motionlefs; it however renewed its ofcillations on being touched.

The confequences of all thefe experiments together may reafonably induce us to fufpect, that the venom of the viper acts on the nerves, and that when it is applied to thefe parts in frogs, it produces a mortal difeafe. But this mode of experimenting is not entirely irreproachable. The fpinal marrow and brain are too fmall to enable us to be certain that the venom does not communicate to the adjacent parts. No precaution whatever can in my opinion prevent this. When the venom is applied, it is too near to the veffels and other parts; and how indeed can it be kept from the blood-veffels of both brain and fpinal marrow?

This

This enquiry is too important to be confined to the limits of fimple probability. I ftill flattered myfelf, that the purfuit of it would tend very much to the obtaining of a true knowledge of the venom of the viper and its qualities, and of the animal economy itfelf.

With this view I formed a plan of experiments, to be made on the nerves of the largeft rabbits I could procure. This animal is hard to kill, and as it is gentle in its nature, may be managed agreeably to one's wifh; it is likewife not fo fmall, but that its nerves may ferve for the moft decifive experiments.

Experiments on the fciatick Nerve of Rabbits.

I made choice of the fciatick nerve as the fubject of my principal experiments. I removed the hair with fciffars from the fkin that covers the great gluteus mufcle, and made an incifion, beginning on the great trochanter, and defcending in a direction with the thigh. I detached the anteriour part of the gluteus mufcle from the os innominatum and trochanter, and gradually raifed the mufcle with my fingers, freeing it from the adipofe membrane. A little cuftom in thefe experiments, enables one to make them in lefs than two minutes, and in fuch a way, that after removing the fmall quantity of blood which flows from the integuments, it is not fucceeded by any frefh hemorrhage that is capable of retarding or difturbing the operation. Now, holding the great gluteus mufcle with one of my hands, I paffed, with the affiftance of fmall pincers, a piece of fine linen in feveral folds under the fciatick nerve, upon which, being in this ftate, I began my experiments.

Having got ready one of the fciatick nerves of a large rabbit, in the way I have juft defcribed, I wounded it in feveral places with a venomous tooth. The rabbit fhook itfelf a little at the time of my doing this. At the end of twenty hours it ate, and feemed in full vigour. It died however at the end of feven days, with a large wound in the part that had been cut. This experiment was not made fo well as it fhould have been ; more than half the gluteus mufcle was cut, and there was a great hemorrhage.

I laid one of the fciatick nerves of another rabbit perfectly bare, paffing beneath it feveral folds of linen. I then wounded it, in upwards of twenty places, with the venomous teeth of two vipers. The rabbit fcarcely gave any figns of feeling pain, and at the end of ten hours, ate and appeared lively. It was in this ftate at the end of twenty-four hours, but died at the end of forty-eight. The nerve was marked here and there with dark red fpots; the parts about it were violently inflamed ; and the blood in the auricles and heart black and coagulated.

In wounding the part with the venomous teeth, I took the greateft care imaginable to prevent the venom from communicating to the adjacent parts ;

2　　　　　　　　　　　　　　　　and

and conftantly covered the nerve after I had wounded it.

Having got ready one of the fciatick nerves of
another rabbit, I paffed under it the ufual folded
linen, and wounded the nerve in feveral places with
the venomous teeth of two vipers. I covered the
nerve well with linen, and ftitched up the fkin. I
had obferved this latter precaution in the preceding
experiments.

The nerve was prepared for the introduction of
the venom in lefs than two minutes, and the hemorrhage from the integuments was very flight indeed. Notwithftanding this the rabbit died at the
end of eighteen hours. The nerve was to appearance in its natural ftate. The blood in the heart
and auricles was black and grumous. The mufcles in the vicinity of the nerve were a little inflamed, and there was a degree of lividnefs on their
fuperficies.

Thefe experiments, although few in number, and
not diftinguifhed by any great uniformity, began
however to make me fufpect, that the bite of the
viper is lefs dangerous to the nerve, than to many
other parts of an animal. The rabbits lived much
longer than one would naturally have conceived,
and notwithftanding it appears that they all died
fooner or later, I conjectured that here, as in the
cafes of the tendons, the venom might have communicated to the neighbouring parts, and that the
animal might probably have died rather from this,
than from any other caufe.

As

As a ftill further precaution in purfuing thefe experiments, I had recourfe to the piece of lead I had before made ufe of, putting it betwixt the folds of the linen. In this way the nerve was very well fecured from the other parts, and it did not feem poffible for the venom to fpread beyond it.

I wounded a fciatick nerve of a rabbit in feveral places, after having got it ready in this way, with the venomous teeth of two vipers; covering it with linen, and binding it up fecurely afterwards. During the time employed in forcing the teeth into the nerve, the rabbit cried out feveral times, and was feized with violent convulfions. It died at the end of twenty hours. All the mufcles about the nerve were livid and fphacelated for their whole fubftance; and the fphacelus extended for the whole length of the leg. The lungs were fpotted; and the nerve itfelf was likewife covered with red and livid fpots. The blood in the auricles and great venous veffels was black and coagulated.

The circumftances that accompany this experiment are fufficient to induce one to believe, that the venom of the viper has in effect a ftrong action on the nerves. The fphacelus of fo many mufcles, even of thofe that were diftant from the wounded part, made a great impreffion on me. I, however, did not on this account terminate my experiments.

Having laid one of the fciatick nerves of another rabbit perfectly bare, I wrapped it carefully in the linen, in which I had not, however, enclofed the bit of lead. I now wounded it in feveral places with the

the teeth of two vipers, and covered it with linen as ufual. The rabbit died at the end of thirty-
two hours. The nerve was but little redder than it
naturally is, and was not fpotted. The blood in the
auricles and large veffels was but flightly coagulated.
When I opened the rabbit, I found it ftill to poffefs
a degree of warmth.

This experiment is very different from the pre-
cedingone, and fhows how little confidence we ought
to place in experiments themfelves, however nicely
they may have been made, unlefs they are in a great
number, and agree with each other.

I laid bare a fciatick nerve of another rabbit, and
wrapped it carefully in the folded linen, in which I
had previoufly enclofed the bit of lead. I wounded
it in feveral places with the venomous teeth of two
vipers, and afterwards covered it fecurely. The
rabbit died at the end of thirty-two hours. Several
parts of the nerve were red, and there were livid
fpots in it. The mufcles adjacent to it were in their
natural ftate; but the lungs were livid and fpotted.
The heart, auricles, and principal veffels, were filled with black grumous blood.

I repeated in four other rabbits, the application
of venom to one of the fciatick nerves, but with
fome little change in the circumftances. I apprehended, that probably the linen which enclofed the
nerve on all fides, and remained on the wound,
might occafion the death of the animal, and the derangements we have obferved. It therefore became
neceffary to feparate thefe two circumftances,

and

and to remove the linen after wounding the nerve with the venomous teeth. Before the linen was removed, I wiped the venom carefully from the surface of the nerve with small brushes, which I repeatedly changed. After this, I dipped bits of linen in water, and holding them with pincers, employed them in washing the whole extent of the nerve. The linen I had passed under the nerve, folded upwards of ten times, prevented the water from communicating to the adjacent parts. I now removed this linen, and threw on the nerve a considerable quantity of water, which washed at once, nerve, muscles, &c. so that it was not possible for any particle of venom, however small, to continue lodged in the parts that surrounded the nerve.

These four rabbits all died in less than thirty-seven hours. In three of them no sensible change was to be observed in the parts adjacent to the venomed nerve. The muscles, except that they were a little redder than usual, were in their natural state.

I confess that on one hand it did not appear possible to me, that the venom could, notwithstanding all the precautions I had taken, have been commucated to the surrounding parts; and on the other hand, I could not find any symptom of disease, any effect of the venom, in the muscles adjacent to the venomed nerve. The death of the animal was the most constant result in these experiments; and this did not however take place till very late, and was not attended with either spasms or convulsions. If

the

the bite of the viper is really venomous to the nerves of animals, it is certain that it acts on thefe parts with lefs force and activity than on many others.

As this enquiry appeared to me a very important one, I thought it proper to perfevere, making fome little change in my experiments.

Experiments on the Sciatick Nerve, divided in its upper Part.

I laid bare the fciatick nerve of a rabbit on one fide, in the ufual way, and with a pair of fciffars divided it in its upper part, as near to the vertebræ as poffible. The cleared part of the fciatick nerve, towards the extremity, was about an inch and an half in length. I wrapped it in linen, which was as ufual in feveral folds, wounded it in feveral places with venomous teeth, and covered it fecurely, to prevent the venom from communicating to the furrounding parts. The rabbit died at the end of thirty-fix hours.

I opened it whilft it was yet warm. The blood in the heart and auricles was black, but not grumous. The mufcles adjacent to the nerve were fomewhat inflamed.

I laid bare one of the fciatick nerves of another rabbit, and divided it in the manner defcribed above. I wrapped it in linen, wounded it with venomous teeth, and covered it. The rabbit died at the end of eighteen hours. The nerve in fome parts was

dark,

dark and livid ; the adjacent mufcles were in a very
flight degree inflamed ; and the blood in the heart
ftill fluid.

The principal aim of this method of experimenting, was to fee what effect the venom of the viper
would produce, when applied immediately to a
nerve, leading it is true to an organized part, and
one endued with fenfation, but which had no longer any immediate communication with the life of
the animal. The action of the venom in the two
above cafes could not in any poffible way be communicated from the nerve to the animal, and could
not awaken in it any immediate pain or fenfation.
This however did not prevent the nerve from communicating the difeafe of the venom to the inferiour
part in which it terminated. It muft be obferved,
that in this part the humours continue in motion as
before ; that the mufcles are in their entire natural
ftate ; that the fibres ftill preferve their irritability ;
and that the part retains its fenfibility, in confequence
of the other nerves that terminate there. With all
this, no difeafe is obferved in it. There is no tumour, no gangrene nor fphacelus, and no extravafation of black and grumous blood.

Thinking, however, that two experiments alone,
would not be fufficient to render this particular,
which is of fo much importance, certain, I was defirous of repeating them in the fame way.

I deftined fix rabbits to this purpofe, and in each
of them laid bare and divided one of the fciatick
nerves, wounding it as ufual with the venomous
teeth,

teeth, and covering it carefully with linen. Thefe rabbits all died, two in eighteen hours, the other four before the expiration of thirty-fix. The adjacent mufcles were in their natural ftate, and the nerves, in a greater or lefs degree, dark and fpotted.

It is therefore certain, that the venom of the viper is not communicated by the nerve to the parts into which that nerve enters and ramifies; notwithftanding it is true that the animal dies on which this experiment is made.

Experiments on the Sciatick Nerve, divided in its inferiour Part.

But if the difeafe of the venom is not communicated to the parts beneath that in which the nerve has been divided, it may neverthelefs be communicated to the upper parts, with which the nerve ftill preferves its former union and correfpondence entire. The animal continues to be fenfible to the fmalleft violence offered to the nerve, which confequently never ceafes to be an organ and inftrument of fenfation, and in which that principle, whatever it may be, ftill exifts, that produces fenfation in the machine.

Having laid bare the fciatick nerve in the ufual way, inftead of dividing it in the upper part towards the vertebræ, I divided it in the inferiour part towards the feet. The cleared part of the nerve was, as ufual, about an inch and an half in length. I

wrapped

wrapped it in linen as in the preceding experiments, and wounded it with the venomous teeth, taking care to cover the whole carefully, to prevent the communication of the venom to the adjacent parts.

The following are the experiments I made in this way :

The fciatick nerve of a rabbit being laid bare, I cut it in the inferiour part towards the feet, and wrapped it in linen which had been folded feven times. I now wounded it repeatedly with the venomous teeth of two vipers, and, in the midft of this operation, the rabbit exhibited ftrong marks of pain. It died at the end of twenty hours. The nerve was fpotted and livid, as were likewife the lungs. The blood in the heart was black and grumous ; but the mufcles about the nerve fcarcely feemed to have undergone the fmalleft change.

This experiment feems to confirm us ftill more in the opinion, that the venom is not communicated to the mufcles adjacent to the nerve, and that there is no local difeafe in thefe parts.

I laid bare the fciatick nerve of another rabbit, cut it in its inferiour part, and wounded it as ufual with the venomous teeth of two vipers. The rabbit cried out and writhed itfelf during the act of wounding the nerve, and died at the end of fixteen hours. The nerve was livid and inflamed in feveral parts. The lungs were covered with large black fpots. The heart, auricles, and large venous veffels, contained black grumous blood. All the adipofe membrane covering the mufcles of the abdomen, was

inflamed,

inflamed, and inner part of the skin was so likewise.
The skin, adipose membrane, and muscles towards
the breast, were all gangrened. The muscles adjacent to the nerve were livid for the depth of a line.

This experiment is very different from the preceding one, and may induce one strongly to suspect,
that the venom of the viper is also venomous to the
nerves, and that in these cases, the disease of the
venom is communicated to all the parts of the animal, above that where the nerve has been cut. In
such an uncertainty there is no way of coming at
the truth, than that of pursuing the experiments.
It is almost impossible not to obtain, in a long continuance of them, some agreement and constancy in
the effects.

I divided one of the sciatick nerves of a rabbit in
the usual way, and having wrapped it in linen,
wounded it with the venomous teeth of two vipers.
The rabbit shrieked violently at the moment of its
being wounded, and died at the end of thirty-seven
hours. The nerve was full of black and livid spots,
and the parts adjacent somewhat inflamed. The
heart was very hard, and very much shrivelled. I
did not open the rabbit till upwards of an hour after its death. The venæ cavæ, however, still oscil-
lated with force. Their motion began at the part
where they open into the auricle, and they continued to move for upwards of five hours longer,
notwithstanding the cavity of the thorax was ex-
posed to the open air.

Having

Having divided the fciatick nerve on one fide, of another rabbit, and wrapped it carefully in linen, I wounded it in feveral places with the venomous teeth of two vipers. The rabbit died at the end of fixteen hours. The nerve had feveral black fpots on its furface, and the adjacent mufcles were livid throughout their whole fubftance. The blood in the heart, auricles, and large venous veffels, was fluid, and a little darker than ufual.

I repeated this experiment with the fame circumftances on fix other rabbits, and the confequences were perfectly analogous to thofe I have related above. The rabbits all died fooner or later, but neither of them in lefs than fixteen hours, nor after thirty-feven. In fome of them the mufcles circumjaeent to the nerve were inflamed and livid throughout their whole fubftance; and in others, on the contrary, they were fimply a little redder than ufual. The blood in the heart, in fome of the cafes, was fluid, and in the others coagulated. The mufcles, adipofe membrane, and fkin, of the breaft, were inflamed in only one inftance. The only thing conftant in thefe experiments was the death of the animal.

We may, in my opinion, deduce in general, from the experiments I have thus far related on the nerves, that the changes obferved in the mufcles adjacent to the fciatick nerve, or thofe in other parts of the animal, are entirely accidental; fince they fometimes exift, and fometimes do not.

Experiments

Experiments on the Sciatick Nerve, on which a Ligature was made.

A new fpecies of experiments remained to be made on the nerves, which might probably decide the queftion. I reflected that the nerve could only communicate the difeafe of the venom to the animal, in confequence of there being a free communication betwixt the nerve and the animal itfelf, and thought of putting an entire ftop to this communication, without even cutting the nerve. We know that a thread which makes a fmall preffure on a nerve, entirely prevents this communication; that the mufcle no longer obeys the will of the animal; and that the nerve is no longer the inftrument or organ, either of motion or fenfation.

In confequence of this hypothefis, I laid bare the fciatick nerve on one fide, of a rabbit, and tied it ftrongly in two parts with a thread. There was a portion of nerve betwixt the two ligatures of more than ten lines. I covered it with a piece of linen in feveral folds, and wounded it repeatedly with the venomous teeth of two vipers, taking care to cover all the parts about it effectually, to prevent a communication of the venom. The rabbit died at the end of fixteen hours. The part of the nerve betwixt the ligatures was white; the mufcles adjacent to the nerve were but very little redder than ufual; the heart, auricles, and great venous veffels,

were

were filled with a fluid blood, fcarcely darker than it is in its natural ftate.

I laid bare the fciatick nerve on one fide, of another rabbit, and tied it in the way I have juft defcribed. I then wounded it betwixt the two ligatures with venomous teeth, and covered it with linen. The rabbit died at the end of eighteen hours. The nerve was in its natural ftate. The adjacent mufcles were red and livid, for the depth of four lines and more.

Having laid bare one of the fciatick nerves of another rabbit, I wounded it as above. The rabbit died at the end of feventeen hours. The nerve was in its natural ftate, and the mufcles about it fcarcely inflamed.

Thefe three experiments fhow, that the greater or lefs degree of inflammation and lividity in the mufcles adjacent to the fciatick nerve, is not owing to the venom ; and that even the death of the animal may arife from fome other caufe. It is very certain, that in the cafes in which the nerve is tied, we do not fee any livid fpots on this part; and, confequently, that they are occafioned by the free communication of the nerve with the animal.

I repeated this experiment, with the fame circumftances, on four other rabbits, all of which died in lefs than nineteen hours. The nerve in each of them was white, and in its natural ftate. The adjacent mufcles in two of them were fcarcely inflamed ; in the other two they were livid for a cer-

tain

tain depth. In one of thefe two laft, a part of the pectoral mufcles was inflamed.

I confefs, that in combining all thefe experiments together, I find nothing that can give me the fmalleft fufpicion, of the nerve being a means of communicating the venom of the viper to an animal, and of exciting in it the difeafe this venom occafions. It is true, that there are livid fpots on the venomed nerve, which are not obferved when it is tied; but may not thefe be purely mechanical, and the effect of the wounds made by the teeth? And even though they fhould be occafioned by the venom itfelf, does it on that account follow, that the venom acts on the nerve, as a venom and not otherwife? Is it therefore demonftrated, that the nerve ought to communicate it to the other parts of the animal?

We are now acquainted with the confequences that enfue from the application of the venom to the fciatick nerve; when this nerve is entire; wh. it is cut, as well above as below; and laftly, when there are two ligatures made in it. It remains to compare all the effects already known, with thofe that will be obferved, in inflicting on the nerve fimple mechanical wounds. After what we have feen, thefe comparative experiments can leave no future doubt.

As the experiments thus far related on the fciatick nerve, were made in three different ways, fo I fhall divide the comparative experiments into three correfponding claffes.

Expe-

Experiments on the Sciatick Nerve, in which mecha-nical Wounds were made.

Having laid bare the fciatick nerve on one fide, of a rabbit, and wrapped it, as ufual, in linen, to the end that all the circumftances might agree with thofe of the preceding experiments, I wounded it in feveral places with a viper's tooth, that had been dried for upwards of a month, and had been carefully wafhed in water, to remove all fufpicion of its con-cealing any venom. The rabbit appeared to fuffer violently when the tooth pierced the nerve. It died at the end of twenty-four hours. The nerve was in feveral parts red and livid; the mufcles ad-jacent to it were inflamed and dark, and thefe ap-pearances extended to the lower part of the leg. The abdominal mufcles and integuments were like-wife inflamed. The right ventricle contained gru-mous blood.

I laid bare one of the fciatick nerves of another rabbit, and having wrapped it in linen as ufual, I pierced it in feveral places with the point of a fine needle. The animal fhrieked terribly, and died at the end of thirty-fix hours. There were feveral dark fpots in the nerve, and the parts adjacent to it were fomewhat inflamed. The blood in the heart was black and coagulated.

Having laid bare one of the fciatick nerves of another rabbit, and wrapped it in linen, I pricked

it

it feveral times with a needle. The rabbit exhibited marks of pain, and died at the end of twenty-feven hours. The mufcles about the nerve were fomewhat livid and inflamed, and the nerve itfelf covered all over with red and black fpots. The blood in the heart was black and coagulated.

Several important truths are demonftrated by thefe experiments.

I. That the livid and red fpots of the nerve are the effect of fimple mechanical wounds.

II. That the death of the rabbits is owing to the fimple wound of the nerve, and not to the venom.

III. That the venom of the viper, communicated to the nerves, neither occafions in any degree the difeafe of the venom, nor haftens the death of the animal.

IV. And laftly, that the venom of the viper is altogether innocent to the nerves, having no greater action on them than pure water, or the fimple folution of gum arabick in diftilled water. I have affured myfelf by other experiments, that it is not at all offenfive to thefe organs.

The experiments I have juft related were not yet fufficient to fatisfy and convince me perfectly. I knew by experience how eafy it is to be mifled by facts, when they are but few in number, and was therefore defirous of repeating the fame procefs over again on four other rabbits. The event was perfectly fimilar to that of the three cafes related above. The rabbits all died; the feiatick nerve, in each of them, was more or lefs covered with livid and red fpots; the adjacent mufcles were, in a greater or lefs degree, inflamed and livid; and the blood in the heart was in general black and coagulated.

Experiments on the Sciatick Nerve.

Having laid bare the fciatick nerve on one fide, of a rabbit, I tied it in two places with a thread, and pricked it feveral times with a needle betwixt the two ligatures. The rabbit died at the end of thirty-three hours. The lungs had feveral dark fpots; the nerve was white, and in its natural ftate; the blood in the heart was dark, but fluid. The rabbit, when I opened it, was ftill warm.

Having laid bare the fciatick nerve on one fide, of a fecond rabbit, and tied it in two places, I pricked it betwixt the two ligatures with a needle. The rabbit died at the end of eighteen hours. The nerve was white, and in its natural ftate; the blood in the heart black and coagulated; and the mufcles that furrounded the nerve red and livid.

I repeated this experiment on two other rabbits, tying the nerve, and pricking it with a needle, as ufual. Both rabbits died; one at the end of thirty hours, the other at the end of thirty-five. The nerves were in a natural ftate, but the mufcles were inflamed, and in one of the rabbits, livid for a confiderable

ſiderable depth. The blood in the heart was black, and grumous.

Experiments on the Sciatick Nerve divided above and below.

Having laid bare the ſciatick nerve of a rabbit, I divided it in its inferiour part, and wrapped it in linen, in the way I had done in all the caſes related above. I pricked it ſeveral times with a needle. The rabbit gave repeated ſhrieks, and died at the end of thirty-ſeven hours. The nerve was covered with black and livid ſpots ; the adjacent parts were ſomewhat inflamed, and the heart ſhrunk, and very hard. The venæ cavæ continued to move for five hours after I had opened the thorax, their motion beginning at where they ariſe from the auricles.

I divided the ſciatick nerve of another rabbit, and having wrapped it in linen, pricked it ſeveral times with the point of a needle. The rabbit died at the end of fifty-four hours. There were black ſpots in ſeveral parts of the nerve; the muſcles about it were ſcarcely inflamed; the blood in the heart was in a fluid ſtate.

I made the ſame experiment on another rabbit, the ſciatick nerve of which, when divided, I pricked ſeveral times with a needle. The rabbit died at the end of thirty hours. The nerve was in ſeveral places red and livid; the muſcles were livid

A a 2 and

and inflamed; and the blood in the heart black and grumous.

I was defirous of repeating the fame experiment, with precifely the fame circumftances, on four other rabbits, all of which died in lefs than forty hours, and one of them before the expiration of eighteen. The mufcles in all of them were in a greater or lefs degree inflamed, and the nerve more or lefs red and livid. The blood in the heart, in fome of them only, was black and coagulated.

Seeing that all thefe experiments correfpond in a certain degree, both with each other, and with the relative ones of the venomed nerves, I did not think it neceffary to make a great number of trials on the fciatick nerve cut in the upper part. I therefore made only two, and thefe agreed in their confequences with thofe of a fimilar kind in which I employed the venom.

I do not conceive that any doubt can remain after thefe experiments, as to the entire innocence of the venom of the viper, applied to the fciatick nerve, and as to the impoffibility of the bite of this animal producing the difeafe of the venom, when confined to a nerve alone.

This new truth in animal phyficks is of the greateft importance in underftanding the nature of the venom of the viper, and its action on the animal body. I muft acknowledge, that I had need of all the experiments on the nerves thus far related, and which are in fo great a number, and varied fo many different ways, to be fully and clearly perfuaded of

this

this circumſtance. Every thing concurred to a belief of the contrary. The rapidity of the diſeaſe,
the ſuddenneſs of the death, the momentaneous
loſs of ſtrength, the very violent convulſions, the
very acute pain, and, in a word, every ſymptom that
characterizes the diſeaſes of the nerves, ſeemed to
exiſt in the animal when the nerve was bit. It is
however certain, that the venom of the viper does
not communicate itſelf to the other parts by the
medium of the nerves, and that the ſubſtance of
them, which cauſes the ſenſation of the animal, and
on which life itſelf ſeems to depend, is not changed
by the action of this venom. The experiments
are direct, they are in a great number, and varied
exceedingly ; the fact is certain, and the errour
was on our ſide, the offspring of prejudice and opinion, and not of nature and experience. On the
other hand, we have ſeen that the venom of the
viper, introduced into the blood, without touching
any veſſel or any ſolid part, kills animals inſtantly,
bringing on very acute pains, and very violent convulſions. I have ſeen the relaxed ſphincters give
paſſage both to the urine and fœces.

Here an occaſion preſents itſelf of examining the
principles and grounds on which this doctrine of
theoretical and practical medicine, that aſcribes diſeaſes to the nerves, and ſubmits ſo many motions
and functions to a nervous principle, is ſupported.
The field is ſo very extenſive, that, although this
diſcuſſion might be very uſeful to the practice of
medicine, I cannot allow myſelf a moment's ſtay

A a 3 in

in it. It will be sufficient for the present to draw this general conclusion—that the usual symptoms of nervous diseases are equivocal and deceptious ; that they may exist without there being any disease of the nerves ; and that a simple change in the blood may be capable of producing all this derangement, and that instantly.

Had the celebrated English physician, Mead, known that a small quantity of the venom of the viper introduced into the blood, kills a large and strong animal almost instantly ; and that this venom is entirely innocent to the nerves ; he certainly would not have had recourse to the animal spirits, and to the nerves, to explain the action of the venom on animals that have been bit. But he was entirely ignorant of these two important truths, which were likewise unknown to all the other physicians of his time.

Mead employed the same principle, that is to say, the nerves and animal spirits, to explain the nature and effects of the other poisons. The nerves are affected every where ; the animal spirits are deranged and in motion every where ; and tumult and nervous agitations are observed every where. He will have this principle applied, not only to the effects of the venom of the viper and the other poisons, but likewise to several other very violent diseases, and amongst others, the plague. This theory is absolutely false as to the venom of the viper, which it seemed to favour the most. I do not believe it any truer as to several other poisons, par-

particularly thofe of the animal kingdom ; and after the experiments I have made, I do not find it demonftrated as to the plague, and other difeafes.

When we examine the reafons that have determined naturalifts and phyficians to recur to the nerves in explaining thefe difeafes, (whether natural or the effect of poifon) we find them to be founded on two principles—the rapidity of the difeafe itfelf; and the convulfions and very fudden proftration of ftrength in the animal.

The firft of thefe two reafons is of no weight, fince I have fhown, that a very fmall quantity of the venom of the viper introduced into the blood, kills an animal in a few inftants; and the fecond is neither evident nor certain, fince experiment itfelf has demonftrated to us, that a little of this venom, in the fame circumftances, produces very violent convulfions, and caufes a proftration of ftrength in the animal in a few moments, notwithftanding it only touches the fluid parts of the blood. I do not befides think it difficult to explain the convulfions, without recurring either to the animal fpirits, or to the nervous fyftem. In the firft part of this work, I have mentioned the circumftance of convulfions arifing fimply from a want of equilibrium in the parts, occafioned by the unequal diftribution of blood in the organs, and by the unequal lofs of irritability in the mufcles. I did not know at that time, either that the nerve was not obnoxious to the venom of the viper, or that this venom was mortal when fimply introduced into the blood. This fub-

ject

ject appears to me of sufficient importance to deferve a work apart. In that work the other poisons
might be examined, as I have done the venom of
the viper; their effects might be analized; and all
the circumstances attending them might be noted.
What lights might not such an undertaking throw
on animal physicks, on the theory of poisons, and
on medicine itself ! The road is open, and nothing
is needed but a patient and industrious obferver. I
can venture to promise him, in the course of a few
years, the moft brilliant, and perhaps too, the moft
useful, discoveries.

But let us return to our experiments.

Although I had affured myself of the innocence
of the venom of the viper, when applied immediately to the nerves, I had a conftant fufpicion
however, that the nerves might at leaft be neceffary
to the action of this venom on the blood. They
might perhaps depofit fome unknown principle, fome
fubtile fluid, in the veffels of the animal, and this
principle, or fluid, might be effential to life, and
likewife to the found ftate of the blood. This was
a new mode of confidering the action of the venom
on the living body, differing effentially from all the
hypothefes that phyficians had hitherto invented;
and it remained to know whether the venom of the
viper would produce a difeafe of greater or lefs violence, on the nervous communication betwixt the
part bitten and the animal being ftopped.

Bite

To obtain fome information on thefe refearches, I made the following experiments.

I cut off the head of a frog, the leg of which I had bit, twice by a viper. This was followed by no fymptom of difeafe in the part.

I cut off the head of another frog, and, after ftripping off the fkin, had one of its legs bit repeatedly by two vipers. The frog has no fymptom of difeafe.

I cut off the head of a third frog, which I had bit repeatedly in the leg, the fkin ftill remaining on. There appeared fome flight fymptoms of difeafe in the leg. At the end of two hours I introduced a pin into the fpinal marrow, and there was fome fmall degree of motion in the mufcles.

I repeated this experiment on four other frogs that had been deprived of the head. Three of them had no fign of difeafe ; in the fourth there were fome little indications of it.

Thefe experiments appearing to me neither fufficiently clear nor uniform, I was defirous of repeating them on twenty-four other frogs, in which I cut off the head, as ufual. Twelve were bit in the leg repeatedly by feveral vipers, and the twelve others had punctures made in the leg, either with fine needles, or with vipers' teeth dried, and free

from

from venom. The consequences were very vague. Of the twelve that were bit, three only had the difeafe of the venom ; and of the twelve that were not bit, but pricked with needles, or wounded with venomous teeth, one had fymptoms of inflammation and lividity in the leg, that might have been confounded with the fymptoms that accompany the difeafe of the venom.

We may, I think, fay in general terms, that a frog deprived of its head, contracts the difeafe of the venom with greater difficulty on that account; and that in thefe circumftances the part bitten is lefs changed by the venom than otherwife. Thefe experiments, however, did not throw a fufficient light on the queftions I propofed to myfelf to clear up, and I therefore determined to fall on a new mode of experimenting.

Experiments on Frogs, in which the Spinal Marrow was divided.

I divided the fpinal marrow of a frog, at two lines above the part where the nerves that are fent to the legs and feet, rife out of the vertebræ. I then had one of the legs of the animal bit repeatedly by two vipers. There was no appearance of a communication of difeafe.

I repeated this experiment on four other frogs, and the event was the fame; I could not obferve any fign of difeafe in the legs that been bit.

After

After having feparated the head from four other frogs, I entirely deftroyed the fpinal marrow with a bit of wood. I then had the legs of the animals bit, but could obferve no fubfequent fymptom of difeafe.

On repeating the experiment of the divided fpinal marrow on fix other frogs, I found in four of them not the fmalleft appearance of difeafe. There was fome doubt as to the fifth, but the fixth appeared to have a real attack of the difeafe of the venom.

This laft cafe made me fufpect the experiments I have related above on the nerves of frogs, and I therefore thought it expedient to proceed to fome experiment that would be more luminous and lefs equivocal. I procured large rabbits for this pur-pofe.

Bite of the Viper on the Parts, the Nerves of which had been divided.

I divided the fciatick and crural nerves of the right leg of a rabbit. I ftitched up the incifed fkin, and had the leg bit by three vipers, each viper biting thrice. The rabbit returned to its food a little time after it had been bit, and reco-vered. At the end of twenty days I employed it in another experiment. I muft remark, that a degree of motion ftill continued in the leg, and that I had

my

my doubts as to the crural nerve having been effectually divided.

I cut the fciatick and crural nerves of the right leg of another rabbit, affuring myfelf that they were completely divided. Having ftitched up the fkin, I had the leg bit by three vipers, each of which bit thrice. The rabbit died at the end of eighteen hours; the mufcles of the leg that had been bit were black, livid, fwelled, and fphacelated; the abdominal mufcles were inflamed; as was alfo the whole of the internal part of the fkin.

Thefe two experiments could not have been more contradictory to each other; it is however certain, that in the fecond cafe there was a true difeafe of the venom. The firft cafe proves nothing more, than that an animal may, on fome particular occafion, be bit repeatedly, even by feveral vipers, and not have an attack of the difeafe. Cafes analogous to this have been related in the preceding parts of this work.

I cut the fciatick and crural nerves of the leg of another rabbit; the nerves were effectually divided, and the leg had loft all motion. I then had the part bit by three vipers, repeatedly by each. The rabbit died at the end of fixteen hours. The mufcles of the leg were livid and gangrened throughout their whole fubftance.

I repeated this experiment with the fame circumftances on two other rabbits, one of which died at the end of twenty hours, the other at the end of twenty-four. They had both the moft determined ed fymptoms of the difeafe of the venom, in the leg that had been bit.

Thefe experiments are pofitive and uniform, and prove indubitably, that it is of no confequence as to the difeafe of the venom, whether the nerves of the parts that are bit, are, or are not, cut; or whether they communicate, or have no longer any communication, with the animal.

But in thefe experiments, there ftill fubfifts a nervous communication betwixt the part bitten and the animal. This communication is formed by the fkin of the animal that covers the part where the venom is received. We muft therefore put a ftop likewife to this communication, by removing the fkin.

After having divided the fciatick and crural nerves on one fide, of a rabbit, and ftitched the incifed fkin, I made a circular incifion in the leg, at four fingers diftance above the part where I had propofed to have it bit by vipers. As foon as the incifion was made, I ftitched it all round. I then had the leg bit repeatedly by three vipers, the teeth penetrating into the fkin. At the end of two hours there was no fymptom of difeafe; at the end of fix the part that had been bit was vifibly inflamed; at the end of ten, there was an oozing of blood from the part of the fkin that had been punctured by the teeth of the viper; at the end of twenty-two, the blood flowed in ftill greater abundance; at the end of twenty-four, the part was very much fwelled, but no lividnefs was to be difcerned; at the end of

thirty

thirty the skin opened, and a wound formed. The rabbit after living eight days, was employed on another occasion.

In this last experiment, there can be no doubt but that the disease of the venom was communicated to the part bitten, although it was not very violent.

I now thought of making a comparative experiment.

For this purpose I had a rabbit bit repeatedly in the leg by three vipers, without cutting either the nerves or the skin. At the end of eight hours the leg was swelled, but not livid. At the end of twenty-two, a cyst or bladder, filled with a dark-coloured humour, formed betwixt the legs, near the part that had been bit. The rabbit died at the end of forty hours. The skin was broken, and corroded at the part bitten; the muscles of the leg were livid and gangrened; the heart, auricles, and large vessels, were filled with black and grumous blood; and there were concretions of blood even in the aorta, which is usually found empty.

I repeated the preceding experiment on three other rabbits, and, after dividing the nerves, making the circular incision in the skin, and stitching it, had each of them bit repeatedly by three vipers. They all had symptoms of the disease of the venom in the part bitten.

I had another suspicion, that there might still subsist, after all, some nervous communication betwixt the animal and the leg, after the sciatick nerve was divided. I conjectured that the viper
might

might have ftruck its teeth on fome fibre of the great gluteus mufcle, which defcends very low in the leg. This was a fufficient inducement to me to make the following experiments.

I divided the fciatick and crural nerves on one fide, of a rabbit, and made a circular incifion in the fkin, which I afterwards ftitched. I had the leg bit by three vipers; by each repeatedly, but in fo low a part as to avoid the above-mentioned mufcle. At the end of two hours the part that had been bit began to inflame; at the end of twenty-two the fkin was broken, but not puffed up; at the end of forty-two the animal was apparently recovered; and at the end of eight days, was employed in other purpofes.

I divided the fciatick and crural nerves on one fide, of another rabbit, and made an incifion, which I afterwards ftitched, for the whole circumference of the leg. I had the leg bit in its inferiour part by three vipers, each viper biting thrice. At the end of eight hours, the fkin burft, and difcharged a humour; at the end of twenty-two this broken fkin was fwelled and livid; and at the end of fixty, the rabbit was in a dying ftate. I opened it, and found all the mufcles of the leg gangrened, and almoft the whole of the adipofe membrane that covers the abdominal mufcles full of extravafated blood. The blood in the heart was in a diffolved ftate.

I divided the fciatick and crural nerves of another rabbit, and made an incifion in the fkin, which I ftitched, all round the leg. I had the inferiour part

3   of

of the leg bit repeatedly by three vipers. At the end of two hours the part that had been bit was become swelled; at the end of eight it was considerably so; at the end of twenty-two, the skin was broken, but without swelling; at the end of forty-two hours, there was simply a wound at the part bitten. The rabbit lived ten days, and was employed in other experiments.

These experiments show, that the venom of the viper produces its usual effects on the parts bitten, notwithstanding all nervous communication has been stopped, betwixt these and the other parts of the animal. But it is not yet decided, whether, supposing some active principle which mixes with the blood to be separated from the nerve, this principle does, or does not, cease to exist the instant the nerve is divided; particularly as the nerves still remain in the part bitten, although they are no longer the instruments of sensation, and voluntary motion. This reflection made me fall upon the following experiments.

I divided the sciatick and crural nerves on one side, of a rabbit, and likewise made the circular incision in the skin, which I stitched. I left the part in this state during sixteen hours, and then had it bit below the incision by three vipers, each of which bit repeatedly. The rabbit died at the end of twenty-two hours. All the muscles of the leg were livid, gangrened, and putrid; the pericardium was filled with a transparent fluid; the right ventricle and right auricle were filled with black grumous blood;

blood; and the blood in the large veſſels was in
the ſame ſtate.

Effects of the Bite of the Viper on Rabbits, the ſpinal
Marrow of which was divided.

I ſhall conclude my experiments on the parts deprived of their nerves, and bit by the viper, by relating three experiments made on rabbits, the ſpinal
marrow of which had been entirely cut through. I
divided it beneath the loins, and in ſo effectual a
way on all ſides, that there could be no ſuſpicion
of any communication of nerves, betwixt the legs
and the other parts of the animal.

The ſpinal marrow of a rabbit having been divided in the way I have juſt mentioned, and a circular inciſion round the leg made in the ſkin, which I
had ſtitched, I had the inferiour part of the leg bit
by three vipers, by each repeatedly. At the end of
an hour, a ſmall tumour formed in the part bitten ;
at the end of two this part was very much ſwelled,
and very livid; and the rabbit died at the end of
ſeven. The part that had been bit was gangrened
all over, and the gangrene penetrated into the whole
ſubſtance of the muſcles that had been wounded by
the teeth. The blood in the heart was black and
grumous.

I divided the ſpinal marrow of another rabbit,
and by the help of ſciſſars detached a great portion

B b   of

of the fkin that covers the crural mufcles. When the latter were in this manner laid bare, I had them bit repeatedly by three vipers. A few minutes after, there were fymptoms of the difeafe of the venom in the part, and the rabbit died at the end of feven hours. The mufcles that had been bit were livid and inflamed; the blood all about them was extravafated into the cellular membrane; there were livid fpots in the lungs; and the heart was filled with blood, which was almoft in an entire ftate of diffolution.

I repeated this experiment on another rabbit, with the fame circumftances, and the refult was likewife the fame. The rabbit died at the end of fix hours, with the mufcles of its leg affected by the difeafe of the venom.

Thus are we affured, that the nerves which are fent to the parts bitten, in no way contribute to the difeafe of the venom of the viper, and that this venom is altogether innocent to the nerves; important truths we were before ignorant of. What is ftill hidden from our fight, is the occafion of the blood, united with the venom, coagulating in an inftant when it is inclofed in the veffels of the animal, and not coagulating in the open air.

Effects

Effects of the Venom on the Parts of an Animal, the Circulation of which was interrupted.

I hoped to draw fome information from the experiments which follow. They confifted in examining the effects of the bite of the viper, on the parts of animals in which the arteries and veins had been previoufly tied by ligature. This was altogether a novel enquiry, and it was not at all amifs to know the effects that would be produced in fimilar cafes.

I made a ligature in the belly, on the aorta defcendens and vena cava of a rabbit. Having ftitched the fkin, I had the leg of the rabbit bit by three vipers, by each repeatedly. The animal died at the end of nine hours. The leg was gangrened in the part where the bites had been received, but in no other.

I divided and removed in the belly of a rabbit, the arteries and veins that go to the right leg, and likewife removed a large portion of the fkin of the leg, which I had bit at the part where the mufcles were bare, by three vipers, each of which bit thrice. At the end of an hour there were certain fymptoms of local difeafe. At the end of two hours the leg was livid at the part bitten, but not elfewhere. The heart was filled with black and grumous blood.

I made a ligature, as in the firft experiment, on the arteries and veins within the belly of two rabbits. Each of them was bit repeatedly by three vi-

ers

pers. In one of them the ſkin was in an entire
ſtate; in the other it was cut circularly, as in an
amputation, and ſtitched. They were both dead at
the expiration of twenty hours. There were ſymptoms of diſeaſe in the parts bitten. However, the
diſeaſe was ſlight, and neither deep nor extenſive.
The blood in the heart was black and coagulated.

I divided in the belly, the arteries and veins of
another rabbit, but neglected to have it bit by vipers. It died at the end of ſixteen hours. The
lungs were livid; and the heart, auricles, and great
veſſels, filled with black and grumous blood. This
experiment is a ſtill further demonſtration to us,
that the grumous blood in the heart and neighbouring veſſels is an equivocal ſign, when it is taken
alone, and without being accompanied by others.

I repeated this experiment of making a ligature
in the belly, on the veins and arteries, on three other
rabbits, having each of them bit in the leg by three
vipers. They all three died in leſs than ſeventeen
hours. The muſcles that had been bit were attacked by the diſeaſe of the venom, but not the
adjacent ones. This local diſeaſe was but of little
conſequence.

We may deduce with certainty from theſe experiments, that the venom of the viper produces its
uſual effects, even when the parts bitten no longer
participate of the circulation of the blood in the
animal machine. In theſe caſes we ſee in general,
that the diſeaſe is leſs extenſive and leſs violent, than
when the blood circulates in the part; and this

particular

particular agrees very well with the experiments in which the venom was injected into the jugular vein.

Effects of the Venom on Parts, the Veffels of which were cut.

I was defirous of feeing what would happen to a rabbit, the crural arteries and veins of which were tied, and cut beneath the ligature, feveral hours prior to the leg being bit. In thefe cafes the blood has not only ceafed to circulate in the leg, but has been a long time ftagnant: it may already be changed in a great meafure, may have fuftained a confiderable lofs in its quantity, and may be deprived of a fubtile principle of fome kind or other. The rabbit that I got ready in this way, remained in this ftate during upwards of eight hours. At the end of that time, I had it bit in the leg by three vipers, each of which bit repeatedly, the fkin having been previoufly removed from the part. The rabbit died three hours after. The mufcle at the part where the vipers had bit, was fomewhat darker than in the adjacent parts; but this was fcarcely fenfible.

I cut the crural artery and vein of a rabbit, in the fame way beneath the ligature, and waited ten hours before I had it bit. At the end of twenty hours it was very lively, and I had it again bit repeatedly by three vipers, the leg being previoufly bared of the fkin, in the part bitten. It died fix hours after. The mufcles that had been bit were livid throughout

their

their whole fubftance, but the difeafe was confined to the part on which I operated.

I repeated this experiment on two other rabbits, having them bit each in a leg, without removing the fkin, eight hours after the ligature had been made, and the crural artery and vein cut. I took the precaution to make repeated compreffions on the leg, that the arterial and venous blood might flow out at the opening in the veffels. Both rabbits died in lefs than eleven hours. The flefh where the teeth had entered appeared of a deeper colour than ufual, and this difcoloration penetrated to the depth the teeth had extended to. The other parts were in a natural ftate.

I got ready two other rabbits in the fame way, as a comparative experiment; they were therefore not bit by vipers. They were both dead at the end of feventy-two hours.

It now remained to examine the effects of the venom of the viper, after having tied the arterial and venous veffels feparately.

For this purpofe I made a ligature on the vena cava in the belly of a rabbit, and afterwards made the circular incifion in the fkin of the leg, and ftitched it. I then had the leg bit repeatedly by three vipers. At the end of twenty-four hours, there were fymptoms of the difeafe of the venom in the part bitten. I killed the rabbit in this ftate, and found that the difeafe was circumfcribed to the incifion made in the fkin. The mufcles were livid, and the

the adipofe membrane filled with dark extravafated blood.

I tied in the belly, the vena cava of another rabbit, and had its leg bit repeatedly by three vipers. At the end of two hours there was an extenfion of the fkin at the part that had been bit, but fcarcely any fenfible fwelling; at the end of four hours, a moifture exuded from it; at the end of ten hours the fwelling had encreafed; and the rabbit died at the end of fifteen. The part bitten was livid and gangrened throughout its whole fubftance; the difeafe was, however, entirely confined to the leg.

The confequences of experiments on two other rabbits treated in the fame way, were pretty fimilar.

I tied the aorta in the belly of a rabbit, and had its leg, covered by the fkin, bit repeatedly by three vipers. At the end of fix hours, the fymptoms of difeafe were perceptible, and the rabbit died at the end of fifteen. The leg was fwelled and livid, and the difcoloration penetrated fome depth into the mufcles. The blood was black in the part that had been bit, and was coagulated in the large veffels.

This experiment repeated on two other rabbits was attended with pretty much the fame fuccefs.

I fhall conclude this chapter, by relating in a few words, two experiments made on rabbits, in the belly of which I had divided all the lymphatick veffels I could find, as far as the ductus thoracicus. An hour after this operation, I had both rabbits bit in the legs, covered by the fkin, repeatedly by three

vipers.

vipers. At the end of fix hours the leg in each, exhibited the moft certain marks of the difeafe of the
venom. It was livid and fwelled, and a good deal
of humour oozed from it. Both rabbits died at the
end of eighteen hours, and the mufcles of the leg
in each, were livid throughout their whole fubftance.

Expecting nothing from the continuation of thefe
experiments, and feeing that the ftoppage of the
circulation of the lymph and chyle has no influence
on the ufual effects of the venom of the viper, I did
not think it neceffary to make any further progrefs
in this fubject.

CHAPTER V.

*Effects of the Venom of the Viper on Blood expofed to
the open Air.*

ALTHOUGH the experiments hitherto related
afford us very important information, we are ftill in
the dark as to the circumftance of the blood coagulating when united in the veffels with the venom of
the viper, and not coagulating when blended with
the latter in the open air. I have at leaft conceived that I conftantly diftinguifhed a very fenfible
difference in this fluid, when I had the leg of an
animal

animal bit after its being separated from the animal itself, and when I had it bit, either still adhering in an entire state to the animal, or fastened to it mechanically.

In this uncertainty, I judged it expedient to make an analysis, followed by the experiment of Mead, that relates to the effects of the venom of the viper on the blood drawn from an animal ; and as Mead made his experiment on a small quantity of venom and a large quantity of blood, I determined that my trials should be on much smaller quantities of the latter, that the effects might be more sensible.

I let fall into a small conical glass, three drops of the venom of the viper, and twenty drops of the blood that flowed from the neck of a fowl, into which I had made an incision. I reclined the glass, and shook it circularly for ten seconds, that the venom and blood might mix well together.

At the same time I let fall into a similar glass, twenty drops of the blood of the same fowl, in the same state as the last. I shook the glass as I had done on the preceding occasion, that, the venom excepted, all circumstances might be alike. At the end of two minutes, the blood unmixed with the venom coagulated, and was of a fine vermillion colour. On the contrary, the blood united to the venom was black and fluid, notwithstanding it was a little viscous and compact.

I repeated this experiment, and the event was the same. The venomed blood did not coagulate, and, as before, was of a dark colour ; whilst, on the contrary,

trary, the blood that was not venomed coagulated very foon, and ftill preferved its bright red colour.

I repeated this experiment on the blood of a guineapig, one of the legs of which I had cut off. The venomed blood at the end of twenty-four hours was ftill black and diffolved, whilft the other coagulated in lefs than two minutes, and continued to preferve its red colour. The venomed blood did not harden till it dried by degrees and fplit into fcales, and preferved its black colour till the laft; inftead of which the blood in a pure ftate, ftill continued red, even after it had dried and fplit.

The black colour of the blood that was mixed with the venom, agrees very well with the moft ufual effects of the bite of the viper on animals, and with the effects of the venom introduced into the jugular vein of rabbits. But the other circumftance in thefe latter experiments is altogether fingular and unexpected. Inftead of the venom coagulating the blood, as one would fuppofe it ought to do, it even keeps it from that coagulation which is natural to it in the open air, and preferves it in a conftant ftate of diffolution. Here then the venom not only does not produce its ordinary effect on the blood, namely, that of coagulating it, but produces an altogether contrary effect, namely, that of keeping it in a diffolved ftate, and fo preventing that coagulation which would otherwife take place.

This fingular effect of the venom on blood expofed to the air, feemed to promife fome important difcovery, relative to the action of it that fucceeds

the

the bite of the viper in animals. I reflected, that this bite is entirely innocent to the viper itself, and likewise to many other animals with cold blood, and that it is not mortal to certain animals, such as frogs, in which it does not produce the difeafe of the venom till very late, and with difficulty. In confequence of this, I perfuaded myfelf that the effects of the venom on the blood of vipers and frogs muft be very different from thofe it produces on the blood of warm animals, and that from this difference, that of the difeafe and death of thefe animals muft precifely depend. Thefe were my reafonings, and I drew no fmall expectations from them.

I accordingly put into a fmall conical glafs, three drops of venom, and thirty drops of the blood that flowed from the neck of a viper, after I had cut off its head. I fhook the glafs, as ufual. The blood, which became fomewhat darker, did not coagulate. At the end of two hours there was a feparation of ferum, which floated on the top, the red part of the blood being beneath. This was dark, and vifcid like glue, but was not coagulated.

I had at the fame time got ready a comparative experiment. I had put into a glafs of the fame kind, thirty drops of the blood of the fame viper, but without having introduced the venom. I fhook the glafs, as ufual. The blood did not coagulate, and became covered with a great deal of ferum, through which the fanguineous fibres, which were very red, were diftinguifhed. At the end of two

hours

hours the ferum was in a much larger quantity than in the preceding experiment ; at the end of twenty-four the red fibres were in their ufual ftate ; but notwithftanding this, the blood was thinner than that which had been venomed. At the end of thirty-five it was ftill fluid, with a great deal of ferum fwimming on the furface ; at the end of fifty it was become thicker and more tenacious ; and at the end of fixty was dry and red.

I mixed in a glafs, three drops of venom with fifty of the frefh blood of a viper, and received in another glafs fifty drops of the fame blood, without making any addition to it. I fhook both glaffes a little, and equally. The blood unmixed with venom, remained till the laft redder than the other, and with a greater proportion of ferum. At the end of thirty hours the venomed blood coagulated, but the other did not.

We fee by thefe experiments, that the colour of the blood of vipers which has been united with the venom of the animal, agrees very well with that of the blood of warm animals blended with the venom in the fame way, notwithftanding there is a great difference in all the other circumftances. Thefe experiments, however, are as yet too little varied to admit any certain conclufions to be drawn from them.

I put three drops of venom into a glafs, and added thirty drops of the blood of a frog, the head of which I had juft cut off. I likewife put thirty drops of this blood into a glafs, without adding any venom.

venom. I fhook the two glaffes, as ufual, and at the end of thirty minutes examined the blood in each. I found the venomed blood black, but not coagulated. The blood unmixed with venom had lefs ferum than the other, was redder and more fibrous, and was likewife in a fluid ftate. At the end of three hours the blood was black and dif-folved, but vifcous, and without any apparent fe-rum. The other blood had a great deal of ferum on its furface. At the bottom it was red and coa-gulated, but the coagulum was moveable, fibrous, and vifcous.

Not fatisfied with this laft experiment, which I repeated twice with pretty much the fame fuccefs, I determined to make at the fame time, experi-ments on the blood of vipers, on that of frogs, and on that of guineapigs, and to follow minutely all the changes that I fhould obferve.

I took fix conical glaffes, fimilar to thofe I had employed before, and put into each of the three firft, four drops of venom, and thirty drops of blood. In the firft glafs was the blood of a viper, in the fecond that of a frog, and in the third that of a guineapig. In each of the other three glaffes I fimply put fifty drops of blood, taken from one of the fame animals, and in the fame order. I fhook the fix glaffes a little, and equally, and then left them in an undifturbed ftate for fome time. At the end of a few minutes the blood in the three ve-nomed glaffes was black, and much more difco-loured than that in the three others, and was al-

ready

ready coagulated; the blood of the viper unmixed
with venom, was however much lefs fo than the
others, and was perhaps rather vifcous than effect-
ally coagulated. The blood of the viper is befides
of a darker red than either that of the frog or guineapig. At the end of fome time, I obferved that
the venomed blood of the viper, and that of the
frog, had a good deal of ferum on their furface,
but that there was none on that of the venomed
blood of the guineapig. There was likewife no
appearance of ferum on the furface of the blood in
the three glaffes, in which it was unmixed with
venom. At the end of eight hours the blood of
the frog contained as much ferum as the venomed
blood, and was in the fame diffolved ftate as the
laft, but redder. The pure blood of the viper
never had any ferum on the top, and continued
coagulated as ufual; but the venomed blood of
the viper was darker than the other, diffolved, and
extremely vifcous. At the end of three days the
venomed blood of the viper ftill preferved the large
quantity of ferum it had at the beginning; it was,
however, black and vifcous. The blood of the
viper, unmixed with venom, had a little ferum,
was red, fibrous, and almoft wholly coagulated.
The venomed blood of the frog was entirely diffolved, of a greenifh hue, and contained a little
ferum; but that which had not been venomed,
had a great deal of ferum on its furface, was coagulated, and redder than the other. The venomed
blood

blood of the guineapig was black, vifcous, and without ferum.

At the end of eight hours I examined the red globules of the blood in the three venomed glaffes, and found them but very little changed in fhape, and fcarcely different from the globules of the pure blood in the three other glaffes. At the end of eight days, however, I found that the globules of the venomed blood of the viper were confiderably altered in fhape; feveral of them were broken, and they were in general much more changed than thofe of the viper's blood unmixed with venom. The globules of the venomed blood of the frog were al-moft all diffolved; thofe that were not were disfi-gured, and very much broken. The venomed blood of the guineapig had, on the contrary, its globules enlarged; they had in fome meafure changed their fhape, and were in a greater or lefs degree diffolved; they did not differ a great deal from thofe of the pure blood of the fame animal.

Thefe laft obfervations on the globules of blood can be of no ufe in explaining the immediate ef-fects of the venom of the viper on its being intro-duced into the veins; and the changes we have feen in thefe globules are not obferved till a long time after the action of the venom on the animal. If the animal is fmall, it is dead a long time before any of thefe changes are obferved.

I repeated the experiment on the blood of the viper, frog, and guineapig, twice, and the confe-quences, although they were not in every refpect

3   per-

perfectly fimilar, were very uniform. I have therefore deemed it unneceffary to enter into a de-tail of them.

We fee in a general way, that the venom of the viper changes the blood both of warm and cold animals black; that of the animals on which it acts as a venom, and that of thofe on which it has no fuch action. This very uniformity of change of colour fhows, however, that the venom of the viper does not kill animals, in confequence of the principle that changes the blood with which it is united, black. It would otherwife be a venom to the viper itfelf, which it is not.

But it is not the fame as to the coagulation of the blood. The venom acts little, or not at all, on the blood of the viper, and the trifling variations we have obferved in this refpect are not at all to be attended to. It is otherwife as to the blood of the frog, and ftill more fo as to that of the guineapig. Scarcely is the latter in the glafs, when it coagu-lates; inftead of which if it is blended with a few drops of venom, it coagulates no more, and re-mains black, vifcous, and without ferum. This effect of the venom is the more fingular, as one would fuppofe that it ought to produce a very con-trary one. Does the venom, when united with the blood, becaufe it is a venom, or from fome other principle, deprive it of the power of coagulating?

It has been feen that the venom of the viper produces a fenfible change on the blood drawn from the veffels of an animal. In thefe cafes the

blood

blood becomes black, and remains fluid, inſtead of
coagulating, as it conſtantly does, when it is not
mixed with the venom.   On the contrary, this
venom, when it is introduced into the circulation
of an animal, ſuddenly impedes it, by cauſing a
coagulation of the blood.   The effects of the venom on the blood of animals are certain, but we
do not on that account know, on what they depend, nor by what mechaniſm all theſe changes
are wrought.   Does the venom of the viper act on
the blood ſimply as a venom ? that is to ſay, by the
very principle that renders it mortal?—We have ſeen
that this venom is a true gummy ſubſtance, and
that it has all the properties which characterize
gums.   We have likewiſe ſeen, that gums are entirely innocent to animals ; and I have obſerved,
that when they are injected in a very ſmall quantity
into the blood, the death of the animal is by no
means ſubſequent.   But why may not the black
colour of the venomed blood, and the fluidity it
preſerves out of the veſſels, depend on the gummy
principle of the venom ?   We know that gums'
abound with phlogiſton, and that phlogiſton gives a
black tinge to the blood.   It is true, that as a
gummy ſubſtance, it ſeems that it ought rather to
coagulate the blood, than to keep it in a ſtate of
diſſolution ; but experiment alone can reply to all
theſe doubts.

I diſſolved a few grains of gum arabick in a ſmall quantity of warm, diſtilled, water. The mucilage was tranſparent, and almoſt fluid. I put three drops of this mucilage into a glaſs, and added ſixty drops of the blood of a pigeon, at the juncture of its flowing from the divided veſſels.

At the ſame time I put three drops of the venom of the viper into another glaſs, and added ſixty drops of the ſame blood, perfectly in the ſame ſtate.

I ſhook both glaſſes for a minute, that their contents might be well blended together. At the end of two minutes the blood united with the gum and coagulated, its colour continuing red as in its natural ſtate, and no ſeparation of ſerum followed, although I kept it two days in the glaſs. The blood in the other glaſs ſuddenly became black, and continued fluid as uſual.

We ſee by this experiment, that gummy ſubſtances do not give a dark tinge to the blood, and that they have not the property of keeping it in a diſſolved ſtate, and preventing its natural coagulation. Thus, then, the changes that are wrought in the blood by the venom of the viper, are not the effect of a gummy principle, but of ſome other principle yet unknown to us, probably the very one that conſtitutes it a venom ; and indeed we have

hitherto

hitherto been able to diftinguifh nothing more in this humour, than a gummy principle, and a ve-nomous principle deftruĉtive of animal life.

I was afterwards defirous of trying whether the venom of the viper would be no longer a venom when mixed with the blood. For this purpofe, I put into a glafs thirty drops of the blood of a pi-geon warm from the veffels, and three drops o venom. I blended the liquors well together, and after twenty-four hours were elapfed, applied feve-ral drops of the mixture to the mufcles of a pi-geon. The pigeon furvived, and at the end of thirty hours feemed fcarcely to have any fymptom of difeafe.

I prepared a mixture of venom and blood in the fame way, in another glafs, but made it with an equal quantity of each fluid. Two minutes after, I covered with it the wounded mufcles of a pigeon, which, although it furvived, had certain fymptoms of the difeafe of the venom.

I repeated this experiment on four other pigeons. Three of them died in lefs than eighteen minutes; and the fourth had fo violent an attack of the dif-eafe, as not to recover till the end of the fixth day. I got ready two other pigeons in the fame way, and did not employ the venom till half an hour after its union with the blood in the glafs. Both pigeons died.

It appears from all thefe experiments, that the venom does not lofe its deadly qualities in confe-quence of being united with the blood.

We have feen that the venom of the viper is a true gum, and that it has all the effential properties of a gum. Why may not this venom prevent the coagulation of the blood in warm animals, and likewife in feveral of the cold ones, as a fimple gum, and not as a venom? And why likewife may not the blood of the viper be different from that of other animals, fince we fee that the venom is innocent to the viper, and not fo to the others?

In this cafe again it belonged to experiment to decide.

As it did not appear to me that the experiments I have hitherto related were fufficient to explain this difficult phenomenon, of blood which coagulates in the enclofed veffels of an animal, and which does not coagulate in glaffes in the open air, I thought it neceffary to examine more attentively than before, the effects of the venom on the legs of animals cut off, and bit by the viper; and likewife on thofe on which a ligature had been made previoufly to their being bit. I was apprehenfive that I might have made fome miftake in my prior experiments, and that fome neceffary-attention might have efcaped me. It was natural to conceive, that after all I had feen in my laft experiments, I was better prepared for nice obfervance than before.

With this prefumption, I made the following experiments.

I had a pigeon bit in the leg repeatedly by a viper, and a few feconds after cut off the limb. There was a degree of lividity at the precife fpot

wher

where the teeth had penetrated; but this was
ícarcely perceptible.

On repeating this experiment with the fame
circumftances, the confequences were perfectly
fimilar.

I had the leg of another pigeon bit by a viper,
a moment after I had cut it off. There was no
fymptom of difeafe, nor any livid appearance.

I wounded the leg of a pigeon with a venomous
tooth, and cut it off immediately after. There
was fome appearance of grumous blood in the
mufcle the tooth had pierced.

I wounded the leg of a pigeon with a viper's
tooth that had been dry a long time, and at the
fame time wounded the pigeon's other leg with a
venomous tooth. The wounds made with the venomous tooth became livid, and this lividity penetrated into the whole fubftance of the mufcle.
There was no appearance of any change in the
part of the other leg that had been pierced by
the dried tooth.

I wounded the leg of a pigeon with venomous
teeth, and cut it off immediately after. There
was a dark fpot, which however I could fcarcely
diftinguifh, at the part the teeth had penetrated.

I forced the venomous tooth into the leg of a pigeon, and cut it off immediately after. There was
no fymptom of the difeafe of the venom.

I cut off the leg of another pigeon, and wounded
it immediately after with a venomous tooth. There
was a flight appearance of dark extravafated blood.

I agaiø

I again forced a venomous tooth into the leg of a pigeon, and cut it off immediately after. This was not fucceeded by any fymptom of the difeafe of the venom.

I pricked the leg of a pigeon repeatedly with the point of a needle, and cut it off immediately after. I obferved dark and extravafated blood at the part that had been pricked.

Although the greater part of thefe experiments fhow, that the venom of the viper has no action on the parts of animals that have been cut off, there are fome of them, however, in which we find a flight appearance of dark and extravafated blood.

The experiment made with the needle rendered what I was defirous of deducing from the others ftill more uncertain. One would conceive that every time the large veffels are ruptured, and there is a fenfible hemorrhage from them, the fpots and dark colour may appear, without the intervention of venom.

It is in general true, that there exifts a notable difference betwixt the effects of the venom of the viper introduced into a leg that has been cut off, and the effects of the fame venom communicated to a leg that continues to make a part of the animal. This difference may occur, either becaufe the quantity of blood in the amputated leg is leffened, becaufe the blood receives fomething from the air, or becaufe it, on the contrary, lofes fomething when in contact with the air. To difcover which

of

•f thefe hypothefes is the moft probable one, I
made the following experiments.

*Experiments on the Venom of the Viper on Limbs that
were fheltered from the Air.*

I placed a pigeon in water in fuch a way, that I
could cut off one of its legs without there being any
communication betwixt the divided part and the air.
A moment before I cut it off I had wounded the leg
with a venomous tooth . At the end of four minutes
I drew it out of the water.   At the part where the
tooth had pierced the mufcle, there was a fmall livid fpot, on which I immediately made an incifion,
and found it to have penetrated juft as far as the
tooth and venom had reached.

I repeated this experiment twice, and the confequence was the fame. . The livid fpot had extended in the fame way into the fubftance of the
mufcle.

The blood of the leg amputated in the water,
flows from the veffels in the fame way as if the
part had been cut off in the open air. The fymptoms, therefore, of the difeafe of the venom in the
leg ftill adhering to the animal, and the abfence of
thefe fymptoms when it is detached, do not depend
on the different quantity of blood that is found in
the two different ftates of the leg.

This experiment feems likewife to determine
that the blood meets with no effential lofs when it

C c 4   is

is expofed to the air, fince it does not feem probable that the water which fuffers the blood to flow
from the leg, prevents this fuppofed principle from
efcaping with it.

It remains probable, then, that the contact of the
air caufes fuch a change in the blood of the leg,
and that the air itfelf unites in fuch a way with it,
as to produce the diverfity of effects we have obferved, although it is certain that we cannot explain what this change confifts in, and how, in
thefe cafes, the air unites itfelf with the blood.

*New Experiments on Parts that were cut, after the
Circulation had been interrupted in them by a Ligature.*

I had now an important experiment to make,
to determine the effects of the venom of the viper
on parts of animals tied, and afterwards cut off.

I had the leg of a pigeon bit by a viper at the
fame inftant that I had it tied and cut off. The
whole operation was made in three feconds, and required the affiftance of three perfons. The amputation was made over the ligature, which was
fo very tight as to prevent the fmalleft hemorrhage.
The leg when cut off had the moft decided fymptoms of the difeafe of the venom. There were
livid fpots on it, the veffels were black and fwelled,
and the blood black, and partly condenfed. Having
opened the mufcles, I found that the livid colour
exten ded

extended for the whole depth of thofe that were wounded.

I immediately made a fecond experiment, fimilar in every refpect to the foregoing, except that the leg was not bit by a viper. It had not the fmalleft fymptom of difeafe.

I had the leg of another pigeon bit once by a viper, and, after four feconds were elapfed, made the ligature and the amputation at one and the fame inftant. In lefs than a minute the fymptoms of the difeafe were apparent, the whole fubftance of the wounded mufcles being in a livid ftate.

I tied and cut off the leg of a pigeon, and immediately after had it bit once by a viper. The fymptoms of the difeafe of the venom in the part were very violent, and the mufcles were livid throughout their whole fubftance.

I tied and cut off the leg of another pigeon, and had it afterwards bit by a viper. The whole fubftance of the mufcles was livid.

Thefe experiments feemed to me fufficiently uniform to difpenfe with my multiplying them any further. They demonftrate, that the venom of the viper acts as a venom on thefe parts, although they are detached from the animals, provided there is no hemorrhage from them.

We likewife fee, that it is not neceffary that the ufual circulation of the blood and other humours ftill continue in the part, fince I have fince obferved that the venom acts on the legs that have been

been tied, even when they have been bit a pretty
confiderable time after the ligature has been made.

*Experiments o. n Animals with warm Blood, deprived of
the Head.*

The experiments made on the frogs that had
been deprived of the head, in which it appeared to
me that the difeafe of the venom was communicated with difficulty, gave me the idea of feeing
whether it would be the fame with warm animals.
Thefe experiments have fome relation to thofe that
were made on the legs of animals cut off and afterwards bit, and only differ from them in this particular—that here the greater part of the body continues in conjunction with the leg, notwithftanding
the blood flows in a great quantity from the neck,
where the incifion has been made.

I divided the trachea arteria of a fowl, and having
fixed in it the nozzle of a fmall pair of bellows, inftantly cut off the head. I now began to work the
bellows, and at the fame time had the leg of the
fowl bit repeatedly by two vipers. The fowl
lived for upwards of fifteen minutes. There were
deep livid fpots in the leg, at the part the teeth
had penetrated.

I repeated this experiment on two rabbits, and
on a guinea pig. They lived beyond all comparifon longer than the fowl, and their life was not
equivocal, as was eafy to be diftinguifhed by the
volun-

voluntary motions of the parts. It is true, that in thefe laft I prevented the lofs of blood, in a great meafure at leaft, by tying the veffels ; and it is certain that they might have lived much longer, if the total effufion of blood could have been prevented. The fymptoms of the difeafe of the venom were manifeft in all the three, and in all of them the mufcles that had been bit were livid.

This experiment fhows, that the head of warm and perfect animals is not neceffary to life, although it is fo to a continuation of it. In a word, an animal may live very well, although deprived of its head, and may be fenfible to external objects. The pulmonary refpiration, and the circulation of humours in the parts, are fufficient for this effect. This principle of life is ftill fuftained in the animal, which may reafonably be faid to be not altogether dead, but only fo in part.

C H A P T E R VI.

On the Cauſe of the Death of Animals bit by the Viper.

MY experiments on animals in which the nerves were bit by vipers, ſhow that the venom is a ſubſtance perfectly innocent to theſe organs, that it does not occaſion in them any ſenſible change, and that they are not even a means or vehicle of conveying it to the animal. In a word, it appears that the nervous ſyſtem does not concur more to the production of the diſeaſe of the venom, than does the tendon, or any other inſenſible part of the animal. On the other hand, all the experiments on the blood, the injection of venom into the veſſels, and ſo on, conſtantly evince that the action of the venom of the viper is on the blood itſelf. This fluid is alone changed by the venom, and this fluid conveys the venom to the animal, and diſtributes it to its whole body. The action of the venom, and its effects on the blood, are almoſt inſtantaneous. The colour of the latter is ſuddenly changed, and the bright red colour that is natural to it, becomes livid and black. This firſt effect is ſucceeded by a ſecond. The blood coagulates very ſuddenly in the lungs, heart, auricles, liver, and in the large venous veſſels. Sometimes the

the heart ftill continues its ofcillatory motions, notwithftanding the blood it contains is, at leaft in
part, coagulated. At other times, the heart beats
with greater force, as if it wifhed to ftop the principle of coagulation that exifts in the blood.

The coagulation of the blood of animals is certainly the moft remarkable effect of the venom of
the viper, and it is this which ought principally to
occafion the derangements in the vifcera, and in
their functions. But the whole mafs of blood is
not coagulated in the animal, fince a part of it appears in a diffolved ftate. The red and lymphatick
parts alone form the coagulum, the ferous part is
more fluid and diffolved than before. It is certain
at leaft, that the latter is thrown in great abundance
on the venomed parts, and fheds itfelf in great
plenty on the adipofe membrane.

If the coagulated part of the blood is left for fome
time in water, it lofes the black colour it had contracted, depofits the red part, which unites with the
water, and leaves a tenacious, white, fibrous, fubftance, fimilar to the polypus.

The blood, partly coagulated, and partly diffolved, produces a very violent derangement in the
organs of the animal. The part bit by the viper
fwells inftantly, and becomes livid by fucceffive
degrees. The blood in the large veins ftops and
coagulates. Tae ferous part tranfudes into the adipofe membrane, which it entirely fills. The circulation is deranged in the vifcera, diminifhes by
degrees, and at length ceafes. The lungs are the
vifcus

vifcus in which the circulation ceafes fooner than in the other parts. In a moment after the injection of venom into the jugular vein, the blood coagulates in the lungs, the veffels of which are filled and diftended with this humour, in a black condenfed ftate. In a word, the circulation is totally impeded and ftopped, and the animal dies. It is a known fact, that as foon as the circulation is ftopped in an animal with warm blood, death enfues in a few minutes, whatever the principle may be that binds and unites together the circulation and the life, the motion of the fluids and the fenfitive faculty.

It will not be foreign to the purpofe to fpeak here of the animal irritability, or of that property of the mufcular fibres, by which a mufcle contracts on the flighteft touch. We muft conceive this property of the mufcular fibres, as fomething that differs from the nerve, or from fenfation; notwithftanding it is true, that the nerve is the organ of the voluntary motions of the animal, and that when it is touched, it excites irritability in the mufcle. The nerve, in whatever way it is ftimulated, is always motionlefs, and the mufcle continues to contract after it is feparated from the animal; whence it follows, that the nerve is rather the occafion than the caufe of the contraction of the mufcles.

In my work entitled *De Legibus Irritabilitatis nunc primum Sancitis*, printed at Lucca in 1767, I demonftrated that the nervous fluid cannot be the *efficient caufe* of mufcular motion. The arguments I adduced in that work are drawn from the hypothefis,

4

thefis, that the nervous fluid acts agreeably to the laws of fluids in general. If the nervous fluid was different from fluids in general, if it had laws altogether different to theirs, or if it was analogous to electricity, my reafons would be no longer applicable to the prefent cafe.

However this may be, it is certain that the motion of a mufcle feparated from the animal, does in no way depend on the animal, or on the fenfitive principle that refides in it, and that the irritability in the fibres fubfifts from itfelf alone. The irritability of the fibres is therefore diftinct from the fenfibility of the animal, and two things which appear fo different, and which feem to have been feparated by nature, ought no longer to be confounded.

But if this fenfitive principle, which conftitutes the life of the animal, is different from the irritability of the fibres, why, in a part feparated from the animal, may there not fubfift an obfcure fenfation, an imperfect life, relative to the fize and to the nature of the part feparated from the animal, and to the nerves that are found in that part?

In this fuppofition, there is no agreement, no harmony, betwixt the life of the entire animal, and the obfcure fenfation of the part that has been feparated: but I do not fee why, in this cafe, the irritability may not likewife depend on the fenfation of the part. The irritability would then depend on the partial fenfibility, or would be the fame as the latter, that is to fay, it would depend on the fenfibi-

lity

lity of the part cut, and not on the fenfibility of the animal.

But the opinion that an obfcure fenfation of life fubfifts in the parts feparated from animals, is founded on an immenfe number of obfervations and experiments, which I have promifed to give in the fecond volume of my Philofophical Enquiries on Animal Phyficks, *(Recherches Philofophiques fur la Phyfique Animale)* the firft volume of which, in quarto, was printed in Italian at Florence, in 1775. In the mean time I can venture to affert, that I know a very great number of animals, even amongft thofe that are called perfect, that is to fay, that have humours, heart, and vifcera, in which the hypothefis of the continuance of animal fenfation, in parts that have been divided, is verified.

But whatever opinion may be adopted on irritability, it is ftill certain that this property exifts in the mufcular fibres, that it is the principle of all the motions of the animal, and that without it, all would be ftill, the organs would become ufelefs, and the functions would be fufpended.

When I wrote the firft part of the prefent work, I was of opinion that the venom of the viper attacked the irritability in an immediate way, and that the animal died from the lofs of irritability in the fibres. But I did not then know, that the venom of the viper has no action on the nerves, and that when it is introduced into the blood, it kills an animal in a few inftants. This hypothefis ought now to be partly modified. It is not that in effect

the

the irritability is not diminifhed, in the animal that has been bit, and that it is not even deftroyed in a little time; but this is rather an effect than a caufe, and is a confequence of the change caufed in the blood by the venom, rather than an effect of the venom on the mufcular fibres. It fometimes occurs that we fee an animal, at the moment of its being bit, lofe all its voluntary motions, and fcarcely difcover any of the lateft fymptoms of life.

The debility of an animal, after it has been bit, is in general very great; but this fhows equally, that the fenfibility is affected: and as the venom does not act on the nerves, but on the blood, this diminution of ftrength and fenfation, and likewife the diminution of the irritability itfelf, may depend on the blood.

I have had frogs bit in the leg by the viper, and have found upon pricking the crural nerves a little time after the bite, or upon drawing electrical fparks from them, that they had loft but little, if any, of their irritability. It is very true, that this irritability diminifhes with time, and that, on the death of the animal, it is frequently loft altogether; but in thefe cafes, the fenfibility is likewife diminifhed and loft. It is befides certain, that if the crural nerves of the leg that has not been bit are ftimulated, they contract with greater force than thofe of the other; and that they frequently contract ftill, when thofe of the venomed leg have no longer in any degree that property.

VoL. I. D d The

The irritability of the fibres, in animals bit by the viper, diminifhes in proportion as the difeafe is more confiderable, and as it continues a longer time. An animal that dies in a few minutes, preferves in its mufcles more irritability, than one that dies at the end of feveral hours, or of feveral days. The irritability ceafes much later in the heart, ftomach, and inteftines, than in the other parts. It particularly ceafes very late in the inteftines, which continue to move, notwithftanding the animal has been dead fome time. The irritability of the diaphragm, or the motion of the thorax, ceafes much later than that of the other mufcles that depend on the will.

I made all thefe obfervations on animals with warm blood, in which it appeared to me, that the electrical fparks were drawn with greater difficulty from the parts bitten, than from the other parts of the animal. This experiment fucceeds particularly in fowls, in which there is no difficulty in laying the mufcles of the leg bare, and in having them bit.

The diminution of irritability in the mufcular fibres, is occafioned by the changes the venom caufes in the blood. The latter in this ftate, in which it is partly diffolved and partly coagulated, is difpofed to a fpeedy putrefaction, and being pent up in the veffels, diffolves the texture of them, paffes through their coats, and fheds itfelf in the adipofe membrane, corrupting and decompofing whatever it meets with. In animals, the parts that have been bit by the viper, pafs in a fhort time to
. the

the ftrongeft putrefaction, and prefent gargrenes and fphacelations. The fkin is fpeedily corroded and deftroyed; the mufcles black and fœtid; and the adipofe membrane diffolved.

I have known a rabbit die in lefs than three hours, with the mufcles of the leg already gangrened throughout their whole fubftance; they were black and offenfive, and were divided by a knife without any refiftance. In a word, this putrefactive tendency of the mufcles, in animals that have been bit by the viper, cannot be denied, and is occafioned by the change wrought in the blood by the venom.

It is very true, that when the animal dies in a few minutes, there is as yet no actual putrefaction in the folid parts, although the humours have a true tendency to this ftate. The difeafe refides folely in the humours, and the ftoppage of thefe humours in their natural courfe, occafions the death of the animal. Whatever tends to impede the motions in the animal machine, neceffarily tends likewife to deftroy in it the fenfitive principle and life; and we cannot conceive life there, where every thing is in a perfect repofe.

Senfation is an active principle, and neceffarily exprefles an action, and we cannot conceive action without motion. We fay in effect, that an animal is dead, when it is no longer fenfible; and we fay, that it is no longer fenfible, when there are no longer in its organs, the figns, the external motions, that indicate fenfation. The moment thefe motions ceafe, we fay that an animal is dead. This manner

of

of judging is founded on obfervation itfelf. We have feen that when an animal is reduced to this ftate of repofe, it does not return again to life ; and think, we may, on the other hand, reafonably con-clude, that an animal, when it is dead, can no lon-ger revive in any manner whatever. This fecond opinion, if we pay attention to it, actually appears to be derived from the firft, fince, after all, we do not know the principle that conftitutes life and fen-fation in animals; it is, however, contradicted by obfervations and experiments of more modern date.

The obfervation that an animal deprived of mo-tion does not return to life, appears to be combat-ted, as I have faid, by modern obfervations of a quite contrary nature. We have heard of ftrong afphyxies, in which there was no longer any fign of motion. We are likewife told of drowned perfons, who have prefented the fame phenomenon, although death in them was nothing more than apparent. I therefore do not fee, why a certain obfcure motion may not fubfift in the organs of an animal, which may not come within the reach of the evidence of our fenfes. A motion to be infenfible is not the lefs real ; and when a motion fubfifts in an animal, there may ftill fubfift in it a principle of fenfation.

I cannot deny but that, when there no longer fub-fifts any principle of fenfation, the animal is in all phyfical rigour dead ; fince we cannot poffibly have any conception of life, in an animal without fenfa-tion. In the fame way it feems equally clear, that

2   a total

a total repofe in the organs of an animal, ought to caufe this fenfation to ceafe, and confequently to occafion the death of the animal. But is there any method by which we can affure ourfelves of the total immobility of the organs of an animal, in which the humours are ftill in a fluid ftate? I cannot conceive any one. A very fmall motion is entirely imperceptible to us, and we fee only the greater ones. Every thing in nature is in motion, and it is not poffible that a body, or any of its parts whatever, can be found for a fingle inftant in a total and perfect repofe. Perfect repofe is befides repugnant to the general laws of gravity, and to the nature of fluids, which are in a greater or lefs degree penetrated by heat. Hence arifes the difficulty of pronouncing on the death of animals, fince in fhort there may ftill fubfift in them a motion which may be infenfible to us, but which may yet be fufficient to maintain in them an obfcure fenfation, to prevent their being altogether dead, and to enable them to return to life.

The motion of the heart being fufpended, and the refpiration and circulation ftopped in an animal, it is foon reduced to that ftate in which we fay of it that it is dead; notwithftanding that this may probably not always be the cafe, when we believe it to be fo. I know of only two ftates of an animal, that can make us certain of its being really dead. One of thefe is the total putrefaction of its organs; the other, the abfolute deficcation of its humours.

D d 3 The

The firſt renders all animal function impoſſible; the ſecond deſtroys all principle of motion.

The total deſiccation of the fluids and ſolids of an animal, not only forbids the uſe of the organs, but even conveys an abſolute immobility into all the parts. An animal, in this ſtate of a total deſiccation of parts, and of an immobility of its organs, is in my opinion certainly dead, and ought ʻo be ſo in the opinion of every body ; elſe we ſhould be ex-poſed to a capricious and unreaſonable pyrrhoniſm. A fiſh, for example, dried in the ſun, or by artifi-cial heat, during twenty years, ſo as to become hard as wood, might ſtill paſs for being alive. I muſt confeſs that I cannot conceive life without action, nor action without motion, nor organical motion when the organs are dry ; and this ſtate is therefore to me a ſtate of death. The naturaliſt ought not, however, to confound with each other theſe two different ſtates of death, that is to ſay, the putrefac-tion of the parts, and the deſiccation of the organs. In the firſt the animal is dead for ever ; in the ſe-cond, it may yet again return to life. We do not know any power, nature herſelf does not diſcloſe any, that can recompoſe an organ that is deſtroyed, and entirely decompoſed by putrefaction, or by the concuſſions of external bodies. This is what has never yet either been accompliſhed or ſeen. We have therefore every poſſible reaſon, not only to be-lieve an animal that is reduced to this ſtate dead, but likewiſe to believe it dead for ever. But if the animal is ſimply dry, if there is no phyſical diſeaſe

4

in its organs, if the component particles of the different parts ftill preferve their refpective fituations, the animal may in this cafe very well return to life, to which effect it is only neceffary, that the organs are reftored to the ftate they were in when the animal was alive. And why ought not an animal to revive in thefe cafes, provided it has every thing that concurred to make it live before ? Whoever had reafoned in this way a century ago, would have advanced matters both probable and reafonable, but would not have been liftened to, even by philofophers, and would have rifked the paffing, at leaft for an extravagant perfon, or for a vifionary.

But let us return to the animals that die by the bite of the viper.

The blood coagulates in the veffels of an animal bit by the viper, and the animal itfelf is in a ftate of death. The blood, changed by the venom, corrupts and deftroys the organs of animals, and renders the leaft fufpicion of life altogether improbable.

It is true that, in proportion as the circulation of the blood ftops in the veffels, and as the death of the animal approaches, we likewife fee a perceptible diminution of the fenfibility ; but this does not yet demonftrate to us, that the nerves are either changed, or have received an injury.

There may perhaps be fuch an harmony or agreement betwixt the circulation of the blood, the air of the lungs, the principle of fenfation, and the

nerves,

nerves, that the one being removed, the other may diminiſh, although one may not act on the other.

My experiments have demonſtrated, that an animal may loſe its ſenſibility from quite another cauſe than from that of an injury to the nerves ; and I therefore think that any one would reaſon ill, who ſhould ſay that the death of an animal depends on the nervous principle alone, becauſe in proportion as the animal draws towards its death, its ſenſibility is found to diminiſh. The diminution of ſenſibility in the nerves may be a ſecondary effect of the cauſe that kills an animal ; and indeed, if the repoſe, if whatever puts a ſtop to motion in the animal, produces death, it muſt likewiſe be productive of a loſs of ſenſation, which cannot ſubſiſt without motion.

Such is the death of animals with warm blood bit by the viper ; but in cold animals it is not exactly the ſame. Animals with cold blood, ſuch for example as frogs, may live a certain time without the circulation of the blood, and without reſpiration. It is preciſely on this account, that the venom of the viper operates on them with leſs activity than on warm animals, and that they ſurvive much longer than theſe laſt, in proportion to the ſize of their bodies. The action of the venom of the viper is inſenſibly communicated to the whole animal; the muſcles diſpoſe to putrefaction, and the part bitten becomes in a little time livid and gangrenous. The death of the animal then follows, but happens much later than in animals with warm blood, be-

cauſe

cauſe the principle of life is not ſo intimately connected with the circulation of the humours.

Why the circulation is thus cloſely connected
with life in animals with warm blood, and why it is
ſo little ſo in animals that have the blood cold, is a
much nicer enquiry. I propoſe to ſpeak on this
ſubject in a work *On Factitious and Natural Airs,*
(Sur les Airs Factices et Naturels) which I hope to
publiſh very ſoon.

End of the Firſt Volume.

I N D E X

B

Bonguer,

Gills

G

James

I

L

M

N

Nerves,

R

    ƒ.   ———— its

W

www.ingramcontent.com/pod-product-compliance
Lightning Source LLC
Chambersburg PA
CBHW021344210326
41599CB00011B/743